Algorithms for
Reinforcement Learning

Algorithms for Reinforcement Learning
Csaba Szepesvári

ISBN: 978-3-031-00423-0   paperback
ISBN: 978-3-031-01551-9   ebook

DOI 10.1007/978-3-031-01551-9

A Publication in the Springer series
SYNTHESIS LECTURES ON ARTIFICIAL INTELLIGENCE AND MACHINE LEARNING

Lecture #9
Series Editors: Ronald J. Brachman, *Yahoo! Research*
                Thomas Dietterich, *Oregon State University*

Series ISSN
Synthesis Lectures on Artificial Intelligence and Machine Learning
Print 1939-4608   Electronic 1939-4616

Synthesis Lectures on Artificial Intelligence and Machine Learning

Editors
Ronald J. Brachman, *Yahoo! Research*
Thomas Dieterich, *Oregon State University*

Algorithms for Reinforcement Learning
Csaba Szepesvári
2010

Data Integration: The Relational Logic Approach
Michael Genesereth
2010

Markov Logic: An Interface Layer for Artificial Intelligence
Pedro Domingos and Daniel Lowd
2009

Introduction to Semi-Supervised Learning
XiaojinZhu and Andrew B.Goldberg
2009

Action Programming Languages
Michael Thielscher
2008

Representation Discovery using Harmonic Analysis
Sridhar Mahadevan
2008

Essentials of Game Theory: A Concise Multidisciplinary Introduction
Kevin Leyton-Brown and Yoav Shoham
2008

A Concise Introduction to Multiagent Systems and Distributed Artificial Intelligence
Nikos Vlassis
2007

Intelligent Autonomous Robotics: A Robot Soccer Case Study
Peter Stone
2007

Algorithms for
Reinforcement Learning

Csaba Szepesvári
University of Alberta

SYNTHESIS LECTURES ON ARTIFICIAL INTELLIGENCE AND MACHINE LEARNING #9

ABSTRACT

Reinforcement learning is a learning paradigm concerned with learning to control a system so as to maximize a numerical performance measure that expresses a long-term objective. What distinguishes reinforcement learning from supervised learning is that only partial feedback is given to the learner about the learner's predictions. Further, the predictions may have long term effects through influencing the future state of the controlled system. Thus, time plays a special role. The goal in reinforcement learning is to develop efficient learning algorithms, as well as to understand the algorithms' merits and limitations. Reinforcement learning is of great interest because of the large number of practical applications that it can be used to address, ranging from problems in artificial intelligence to operations research or control engineering. In this book, we focus on those algorithms of reinforcement learning that build on the powerful theory of dynamic programming. We give a fairly comprehensive catalog of learning problems, describe the core ideas, note a large number of state of the art algorithms, followed by the discussion of their theoretical properties and limitations.

KEYWORDS

reinforcement learning, Markov Decision Processes, temporal difference learning, stochastic approximation, two-timescale stochastic approximation, Monte-Carlo methods, simulation optimization, function approximation, stochastic gradient methods, least-squares methods, overfitting, bias-variance tradeoff, online learning, active learning, planning, simulation, PAC-learning, Q-learning, actor-critic methods, policy gradient, natural gradient

Contents

Preface

Reinforcement learning (RL) refers to both a learning problem and a subfield of machine learning. As a learning problem, it refers to learning to control a system so as to maximize some numerical value which represents a long-term objective. A typical setting where reinforcement learning operates is shown in Figure 1: A controller receives the controlled system's state and a reward associated with the last state transition. It then calculates an action which is sent back to the system. In response, the system makes a transition to a new state and the cycle is repeated. The problem is to learn a way of controlling the system so as to maximize the total reward. The learning problems differ in the details of how the data is collected and how performance is measured.

In this book, we assume that the system that we wish to control is stochastic. Further, we assume that the measurements available on the system's state are detailed enough so that the controller can avoid reasoning about how to collect information about the state. Problems with these characteristics are best described in the framework of Markovian Decision Processes (MDPs). The standard approach to 'solve' MDPs is to use dynamic programming, which transforms the problem of finding a good controller into the problem of finding a good value function. However, apart from the simplest cases when the MDP has very few states and actions, dynamic programming is infeasible. The RL algorithms that we discuss here can be thought of as a way of turning the infeasible dynamic programming methods into practical algorithms so that they can be applied to large-scale problems.

There are two key ideas that allow RL algorithms to achieve this goal. The first idea is to use samples to compactly represent the dynamics of the control problem. This is important for

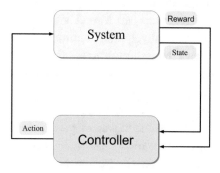

Figure 1: The basic reinforcement learning scenario

two reasons: First, it allows one to deal with learning scenarios when the dynamics is unknown. Second, even if the dynamics is available, exact reasoning that uses it might be intractable on its own. The second key idea behind RL algorithms is to use powerful function approximation methods to compactly represent value functions. The significance of this is that it allows dealing with large, high-dimensional state- and action-spaces. What is more, the two ideas fit nicely together: Samples may be focused on a small subset of the spaces they belong to, which clever function approximation techniques might exploit. It is the understanding of the interplay between dynamic programming, samples and function approximation that is at the heart of designing, analyzing and applying RL algorithms.

The purpose of this book is to allow the reader to have a chance to peek into this beautiful field. However, certainly we are not the first to set out to accomplish this goal. In 1996, Kaelbling et al. have written a nice, compact survey about the approaches and algorithms available at the time (Kaelbling et al., 1996). This was followed by the publication of the book by Bertsekas and Tsitsiklis (1996), which detailed the theoretical foundations. A few years later Sutton and Barto, the 'fathers' of RL, published their book, where they presented their ideas on RL in a very clear and accessible manner (Sutton and Barto, 1998). A more recent and comprehensive overview of the tools and techniques of dynamic programming/optimal control criteria, as well as various classes of controlled systems is given in the two-volume book by Bertsekas (2007a,b) which devotes one chapter to RL methods.[1] At times, when a field is rapidly developing, books can get out of date pretty quickly. In fact, to keep up with the growing body of new results, Bertsekas maintains an online version of his Chapter 6 of Volume II of his book, which, at the time of writing this survey counted as much as 160 pages (Bertsekas, 2010). Other recent books on the subject include the book of Gosavi (2003) who devotes 60 pages to reinforcement learning algorithms in Chapter 9, concentrating on average cost problems, or that of Cao (2007) who focuses on policy gradient methods. Powell (2007) presents the algorithms and ideas from an operations research perspective and emphasizes methods that are capable of handling large control spaces, Chang et al. (2008) focuses on adaptive sampling (i.e., simulation-based performance optimization), while the center of the recent book by Busoniu et al. (2010) is function approximation.

Thus, by no means do RL researchers lack a good body of literature. However, what seems to be missing is a self-contained and yet relatively short summary that can help newcomers to the field to develop a good sense of the state of the art, as well as existing researchers to broaden their overview of the field, an article, similar to that of Kaelbling et al. (1996), but with an updated contents. To fill this gap is the very purpose of this short book.

Having the goal of keeping the text short, we had to make a few, hopefully, not too troubling compromises. The first compromise we made was to present results only for the total expected discounted reward criterion. This choice is motivated by that this is the criterion that is both widely used and the easiest to deal with mathematically. The next compromise is that the background

---

[1]In this book, RL is called neuro-dynamic programming or approximate dynamic programming. The term neuro-dynamic programming stems from the fact that, in many cases, RL algorithms are used with artificial neural networks.

on MDPs and dynamic programming is kept ultra-compact (although an appendix is added that explains these basic results. Apart from these, the book aims to cover a bit of all aspects of RL, up to the level that the reader should be able to understand the whats and hows, as well as to implement the algorithms presented. Naturally, we still had to be selective in what we present. Here, the decision was to focus on the basic algorithms, ideas, as well as the available theory. Special attention was paid to describing the choices of the user, as well as the trade offs that come with these. We tried to be impartial as much as possible, but some personal bias, as usual, surely remained. The pseudocode of almost twenty algorithms was included, hoping that this will make it easier for the practically inclined reader to implement the algorithms described.

The target audience is advanced undergraduate and graduate students, as well as researchers and practitioners who want to get a good overview of the state of the art in RL quickly. Researchers who are already working on RL might also enjoy reading about parts of the RL literature that they are not so familiar with, thus broadening their perspective on RL. The reader is assumed to be familiar with the basics of linear algebra, calculus, and probability theory. In particular, we assume that the reader is familiar with the concepts of random variables, conditional expectations, and Markov chains. It is helpful, but not necessary, for the reader to be familiar with statistical learning theory, as the essential concepts will be explained as needed. In some parts of the book, knowledge of regression techniques of machine learning will be useful.

This book has three parts. In the first part, in Section 1, we provide the necessary background. It is here where the notation is introduced, followed by a short overview of the theory of Markov Decision Processes and the description of the basic dynamic programming algorithms. Readers familiar with MDPs and dynamic programming should skim through this part to familiarize themselves with the notation used. Readers, who are less familiar with MDPs, must spend enough time here before moving on because the rest of the book builds heavily on the results and ideas presented here.

The remaining two parts are devoted to the two basic RL problems (cf. Figure 2), one part devoted to each. In Section 2) the problem of learning to predict values associated with states is studied. We start by explaining the basic ideas for the so-called tabular case when the MDP is small enough so that one can store one value per state in an array allocated in a computer's main memory. The first algorithm explained is $TD(\lambda)$, which can be viewed as the learning analogue to value iteration from dynamic programming. After this, we consider the more challenging situation when there are more states than what fits into a computer's memory. Clearly, in this case, one must compress the table representing the values. Abstractly, this can be done by relying on an appropriate function approximation method. First, we describe how $TD(\lambda)$ can be used in this situation. This is followed by the description of some new gradient based methods (GTD2 and TDC), which can be viewed as improved versions of $TD(\lambda)$ in that they avoid some of the convergence difficulties that $TD(\lambda)$ faces. We then discuss least-squares methods (in particular, $LSTD(\lambda)$ and λ-LSPE) and compare them to the incremental methods described earlier. Finally, we describe choices available for implementing function approximation and the trade offs that these choices come with.

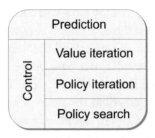

Figure 2: Types of reinforcement problems and approaches.

The second part (Section 3) is devoted to algorithms that are developed for control learning. First, we describe methods whose goal is optimizing online performance. In particular, we describe the "optimism in the face of uncertainty" principle and methods that explore their environment based on this principle. State of the art algorithms are given both for bandit problems and MDPs. The message here is that clever exploration methods make a large difference, but more work is needed to scale up the available methods to large problems. The rest of this section is devoted to methods that aim at developing methods that can be used in large-scale applications. As learning in large-scale MDPs is significantly more difficult than learning when the MDP is small, the goal of learning is relaxed to learning a good enough policy in the limit. First, direct methods are discussed which aim at estimating the optimal action-values directly. These can be viewed as the learning analogue of value iteration of dynamic programming. This is followed by the description of actor-critic methods, which can be thought of as the counterpart of the policy iteration algorithm of dynamic programming. Both methods based on direct policy improvement and policy gradient (i.e., which use parametric policy classes) are presented.

The book is concluded in Section 4, which lists some topics for further exploration.

Csaba Szepesvári
June 2010

Acknowledgments

I am truly indebted to my family for their love, support and patience. Thank you Mom, Beáta, Dávid, Réka, Eszter, Csongor! Special thanks to Réka who has helped me drawing Figure 1.1. A number of individuals have read various versions of the manuscript, full or in parts and helped me to reduce the number of mistakes by sending corrections. They include Dimitri Bertsekas, Gábor Balázs, Bernardo Avila Pires, Warren Powell, Rich Sutton, Nikos Vlassis, Hengshuai Yao and Shimon Whiteson. Thank You! Of course, all the remaining mistakes are mine. If I have left out someone from the above list, this was by no means intentional. If this is the case, please remind me in an e-mail (better yet, send me some comments or suggestions). Independently of whether they have contacted me before or not, readers are encouraged to e-mail me if they find errors, typos or they just think that some topic should have been included (or left out). I plan to periodically update the text and I will try to accommodate all the requests. Finally, I wish to thank Remi Munos and Rich Sutton, my closest collaborators over the last few years, from whom I have learned and continue to learn a lot. I also wish to thank all my students, the members of RLAI group and all researchers of RL who continue to strive to push the boundaries of what we can do with reinforcement learning. This book is made possible by you.

Csaba Szepesvári
June 2010

CHAPTER 1

Markov Decision Processes

The purpose of this section is to introduce the notation that will be used in the subsequent parts and the most essential facts that we will need from the theory of Markov Decision Processes (MDPs) in the rest of the book. Readers familiar with MDPs should skim through this section to familiarize themselves with the notation. Readers unfamiliar with MDPs are suggested to spend enough time with this section to understand the details. Proofs of most of the results (with some simplifications) are included in Appendix A. The reader who is interested in learning more about MDPs is suggested to consult one of the many excellent books on the subject, such as the books of Bertsekas and Shreve (1978), Puterman (1994), or the two-volume book by Bertsekas (2007a,b).

1.1 PRELIMINARIES

We use $\mathbb{N}$ to denote the set of natural numbers: $\mathbb{N} = \{0, 1, 2, \ldots\}$, while $\mathbb{R}$ denotes the set of reals. By a vector v (unless it is transposed, $v^\top$), we mean a column vector. The inner product of two finite-dimensional vectors, $u, v \in \mathbb{R}^d$ is $\langle u, v \rangle = \sum_{i=1}^{d} u_i v_i$. The resulting 2-norm is $\|u\|^2 = \langle u, u \rangle$. The maximum norm for vectors is defined by $\|u\|_\infty = \max_{i=1,\ldots,d} |u_i|$, while for a function $f : \mathcal{X} \to \mathbb{R}$, $\|\dot{\|}_\infty$ is defined by $\|f\|_\infty = \sup_{x \in \mathcal{X}} |f(x)|$. A mapping T between the metric spaces (M_1, d_1), (M_2, d_2) is called Lipschitz with modulus $L \in \mathbb{R}$ if for any $a, b \in M_1, d_2(T(a), T(b)) \leq L\, d_1(a, b)$. If T is Lipschitz with a modulus $L \leq 1$, it is called a non-expansion. If $L < 1$, the mapping is called a contraction. The indicator function of event S will be denoted by $\mathbb{I}_{\{S\}}$ (i.e., $\mathbb{I}_{\{S\}} = 1$ if S holds and $\mathbb{I}_{\{S\}} = 0$, otherwise). If $v = v(\theta, x)$, $\frac{\partial}{\partial \theta} v$ shall denote the partial derivative of v with respect to θ, which, if θ, which is a d-dimensional *row vector* if $\theta \in \mathbb{R}^d$. The total derivative of some expression v with respect to θ will be denoted by $\frac{d}{d\theta} v$ (and will be treated as a row vector). Further, $\nabla_\theta v = (\frac{d}{d\theta} v)^\top$.

If P is a distribution or a probability measure, then $X \sim P$ means that X is a random variable drawn from P.

1.2 MARKOV DECISION PROCESSES

For ease of exposition, we restrict our attention to countable MDPs and the discounted total expected reward criterion. However, under some technical conditions, the results extend to continuous state-action MDPs, too. This also holds true for the results presented in later parts of this book.

A countable MDP is defined as a triplet $\mathcal{M} = (\mathcal{X}, \mathcal{A}, \mathcal{P}_0)$, where $\mathcal{X}$ is the countable non-empty set of states, $\mathcal{A}$ is the countable non-empty set of actions. The *transition probability kernel* $\mathcal{P}_0$ assigns to each state-action pair $(x, a) \in \mathcal{X} \times \mathcal{A}$ a probability measure over $\mathcal{X} \times \mathbb{R}$, which we shall denote by $\mathcal{P}_0(\cdot | x, a)$. The semantics of $\mathcal{P}_0$ is the following: For $U \subset \mathcal{X} \times \mathbb{R}, \mathcal{P}_0(U | x, a)$ gives the probability that the next state and the associated reward belongs to the set U provided that the current state is x and the action taken is a.[1] We also fix a discount factor $0 \leq \gamma \leq 1$ whose role will become clear soon.

The transition probability kernel gives rise to the *state transition probability kernel*, $\mathcal{P}$, which, for any $(x, a, y) \in \mathcal{X} \times \mathcal{A} \times \mathcal{X}$ triplet gives the probability of moving from state x to some other state y provided that action a was chosen in state x:

$$\mathcal{P}(x, a, y) = \mathcal{P}_0(\{y\} \times \mathbb{R} \mid x, a).$$

In addition to $\mathcal{P}$, $\mathcal{P}_0$ also gives rise to *the immediate reward function $r : \mathcal{X} \times \mathcal{A} \to \mathbb{R}$*, which gives the expected immediate reward received when action a is chosen in state x: If $(Y_{(x,a)}, R_{(x,a)}) \sim \mathcal{P}_0(\cdot \mid x, a)$, then

$$r(x, a) = \mathbb{E}\left[R_{(x,a)}\right].$$

In what follows, we shall assume that the rewards are bounded by some quantity $\mathcal{R} > 0$: for any $(x, a) \in \mathcal{X} \times \mathcal{A}, |R_{(x,a)}| \leq \mathcal{R}$ almost surely. It is immediate that if the random rewards are bounded by $\mathcal{R}$ then $\|r\|_\infty = \sup_{(x,a) \in \mathcal{X} \times \mathcal{A}} |r(x, a)| \leq \mathcal{R}$ also holds. An MDP is called *finite* if both $\mathcal{X}$ and $\mathcal{A}$ are finite.

Markov Decision Processes are a tool for modeling sequential decision-making problems where a decision maker interacts with a system in a sequential fashion. Given an MDP $\mathcal{M}$, this interaction happens as follows: Let $t \in \mathbb{N}$ denote the current time (or stage), let $X_t \in \mathcal{X}$ and $A_t \in \mathcal{A}$ denote the random state of the system and the action chosen by the decision maker at time t, respectively. Once the action is selected, it is sent to the system, which makes a transition:

$$(X_{t+1}, R_{t+1}) \sim \mathcal{P}_0(\cdot \mid X_t, A_t). \tag{1.1}$$

In particular, X_{t+1} is random and $\mathbb{P}(X_{t+1} = y | X_t = x, A_t = a) = \mathcal{P}(x, a, y)$ holds for any $x, y \in \mathcal{X}, a \in \mathcal{A}$. Further, $\mathbb{E}[R_{t+1} | X_t, A_t] = r(X_t, A_t)$. The decision maker then observes the next state X_{t+1} and reward R_{t+1}, chooses a new action $A_{t+1} \in \mathcal{A}$ and the process is repeated. The goal of the decision maker is to come up with a way of choosing the actions so as to maximize the expected total discounted reward.

The decision maker can select its actions at any stage based on the observed history. A rule describing the way the actions are selected is called a *behavior*. A behavior of the decision maker and some initial random state X_0 together define a random state-action-reward sequence $((X_t, A_t, R_{t+1}); t \geq 0)$, where (X_{t+1}, R_{t+1}) is connected to (X_t, A_t) by (1.1) and A_t is the action prescribed by the behavior based on the history $X_0, A_0, R_1, \ldots, X_{t-1}, A_{t-1}, R_t, X_t$.[2]

---

[1]The probability $\mathcal{P}_0(U|x, a)$ is defined only when U is a Borel-measurable set. Borel-measurability is a technical notion whose purpose is to prevent some pathologies. The collection of Borel-measurable subsets of $\mathcal{X} \times \mathbb{R}$ include practically all "interesting" subsets $\mathcal{X} \times \mathbb{R}$. In particular, they include subsets of the form $\{x\} \times [a, b]$ and subsets which can be obtained from such subsets by taking their complement, or the union (intersection) of at most countable collections of such sets in a recursive fashion.

[2]Mathematically, a behavior is an infinite sequence of probability kernels $\pi_0, \pi_1, \ldots, \pi_t, \ldots$, where π_t maps histories of length t to a probability distribution over the action space $\mathcal{A}$: $\pi_t = \pi_t(\cdot | x_0, a_0, r_0, \ldots, x_{t-1}, a_{t-1}, r_{t-1}, x_t)$.

The *return* underlying a behavior is defined as the total discounted sum of the rewards incurred:

$$\mathcal{R} = \sum_{t=0}^{\infty} \gamma^t R_{t+1}.$$

Thus, if $\gamma < 1$ then rewards far in the future worth exponentially less than the reward received at the first stage. An MDP when the return is defined by this formula is called a *discounted reward* MDP. When $\gamma = 1$, the MDP is called *undiscounted*.

The goal of the decision-maker is to choose a behavior that maximizes the expected return, irrespectively of how the process is started. Such a maximizing behavior is said to be *optimal*.

Example 1.1   Inventory control with lost sales   Consider the problem of day-to-day control of an inventory of a fixed maximum size in the face of uncertain demand: Every evening, the decision maker must decide about the quantity to be ordered for the next day. In the morning, the ordered quantity arrives with which the inventory is filled up. During the day, some stochastic demand is realized, where the demands are independent with a common fixed distribution, see Figure 1.1. The goal of the inventory manager is to manage the inventory so as to maximize the present monetary value of the expected total future income.

The payoff at time step t is determined as follows: The cost associated with purchasing A_t items is $K\mathbb{I}_{\{A_t>0\}} + cA_t$. Thus, there is a fixed entry cost K of ordering nonzero items and each item must be purchased at a fixed price c. Here $K, c > 0$. In addition, there is a cost of holding an inventory of size $x > 0$. In the simplest case, this cost is proportional to the size of the inventory with proportionality factor $h > 0$. Finally, upon selling z units the manager is paid the monetary amount of pz, where $p > 0$. In order to make the problem interesting, we must have $p > h$, otherwise there is no incentive to order new items.

This problem can be represented as an MDP as follows: Let the state X_t on day $t \geq 0$ be the size of the inventory in the evening of that day. Thus, $\mathcal{X} = \{0, 1, \ldots, M\}$, where $M \in \mathbb{N}$ is the maximum inventory size. The action A_t gives the number of items ordered in the evening of day t. Thus, we can choose $\mathcal{A} = \{0, 1, \ldots, M\}$ since there is no need to consider orders larger than the inventory size. Given X_t and A_t, the size of the next inventory is given by

$$X_{t+1} = ((X_t + A_t) \wedge M - D_{t+1})^+, \tag{1.2}$$

where $a \wedge b$ is a shorthand notation for the minimum of the numbers a, b, $(a)^+ = a \vee 0 = \max(a, 0)$ is the positive part of a, and $D_{t+1} \in \mathbb{N}$ is the demand on the $(t+1)^{\text{th}}$ day. be higher than that of $\wedge$ and $\vee$). By assumption, $(D_t; t > 0)$ is a sequence of independent and identically distributed (*i.i.d.*) integer-valued random variables. The revenue made on day $t + 1$ is

$$\begin{aligned} R_{t+1} = &-K\,\mathbb{I}_{\{A_t>0\}} - c\,((X_t + A_t) \wedge M - X_t)^+ \\ &- h\,X_t \qquad + p\,((X_t + A_t) \wedge M - X_{t+1})^+. \end{aligned} \tag{1.3}$$

Figure 1.1: Illustration of the inventory management problem

Equations (1.2)–(1.3) can be written in the compact form

$$(X_{t+1}, R_{t+1}) = f(X_t, A_t, D_{t+1}), \tag{1.4}$$

with an appropriately chosen function f. Then, $\mathcal{P}_0$ is given by

$$\mathcal{P}_0(U \mid x, a) = \mathbb{P}(f(x, a, D) \in U) = \sum_{d=0}^{\infty} \mathbb{I}_{\{f(x,a,d) \in U\}} \, p_D(d).$$

Here $p_D(\cdot)$ is the probability mass function of the random demands and $D \sim p_D(\cdot)$. This finishes the definition of the MDP underlying the inventory optimization problem.

Inventory control is just one of the many operations research problems that give rise to an MDP. Other problems include optimizing transportation systems, optimizing schedules or production. MDPs arise naturally in many engineering optimal control problems, too, such as the optimal control of chemical, electronic or mechanical systems (the latter class includes the problem of controlling robots). Quite a few information theory problems can also be represented as MDPs (e.g., optimal coding, optimizing channel allocation, or sensor networks). Another important class of problems comes from finance. These include, amongst others, optimal portfolio management and option pricing.

In the case of the inventory control problem, the MDP was conveniently specified by a transition function f (cf., (1.4)). In fact, transition functions are as powerful as transition kernels: any MDP gives rise to some transition function f and any transition function f gives rise to some MDP.

In some problems, not all actions are meaningful in all states. For example, ordering more items than what one has room for in the inventory does not make much sense. However, such meaningless actions (or forbidden actions) can always be remapped to other actions, just like it was done above. In some cases, this is unnatural and leads to a convoluted dynamics. Then, it might be better to introduce an additional mapping which assigns the set of *admissible* actions to each state.

In some MDPs, some states are impossible to leave: If x is such a state, $X_{t+s} = x$ holds almost surely[3] for any $s \geq 1$ provided that $X_t = x$, no matter what actions are selected after time t. By convention, we will assume that no reward is incurred in such *terminal* or *absorbing* states. An MDP with such states is called *episodic*. An *episode* then is the (generally random) time period from the beginning of time until a terminal state is reached. In an episodic MDP, we often consider undiscounted rewards, i.e., when $\gamma = 1$.

Example 1.2 Gambling   A gambler enters a game whereby she may stake any fraction $A_t \in [0, 1]$ of his current wealth $X_t \geq 0$. She wins his stake back and as much more with probability $p \in [0, 1]$, while she loses his stake with probability $1 - p$. Thus, the fortune of the gambler evolves according to

$$X_{t+1} = (1 + S_{t+1} A_t) X_t.$$

Here $(S_t; t \geq 1)$ is a sequence of independent random variables taking values in $\{-1, +1\}$ with $\mathbb{P}(S_{t+1} = 1) = p$. The goal of the gambler is to maximize the probability that his wealth reaches an *a priori* given value $w^* > 0$. It is assumed that the initial wealth is in $[0, w^*]$.

This problem can be represented as an episodic MDP, where the state space is $\mathcal{X} = [0, w^*]$ and the action space is $\mathcal{A} = [0, 1]$.[4] We define

$$X_{t+1} = (1 + S_{t+1} A_t) X_t \wedge w^*, \tag{1.5}$$

when $0 \leq X_t < w^*$ and make w^* a terminal state: $X_{t+1} = X_t$ if $X_t = w^*$. The immediate reward is zero as long as $X_{t+1} < w^*$ and is one when the state reaches w^* for the first time:

$$R_{t+1} = \begin{cases} 1, & X_t < w^* \text{ and } X_{t+1} = w^*; \\ 0, & \text{otherwise.} \end{cases}$$

If we set the discount factor to one, the total reward along any trajectory will be one or zero depending on whether the wealth reaches w^*. Thus, the expected total reward is just the probability that the gambler's fortune reaches w^*.

Based on the two examples presented so far, the reader unfamiliar with MDPs might believe that all MDPs come with handy finite, one-dimensional state- and action-spaces. If only this was true! In fact, in practical applications the state- and action-spaces are often very large, multidimensional spaces. For example, in a robot control application, the dimensionality of the state space can be 3—6 times the number of joints the robot has. An industrial robot's state space might easily be 12—20 dimensional, while the state space of a humanoid robot might easily have 100 dimensions. In a real-world inventory control application, items would have multiple types, the prices and costs

---

[3]"Almost surely" means the same as "with probability one" and is used to refer to the fact that the statement concerned holds everywhere on the probability space with the exception of a set of events with measure zero.
[4]Hence, in this case the state and action spaces are continuous. Notice that our definition of MDPs is general enough to encompass this case, too.

would also change based on the state of the "market", whose state would thus also become part of the MDP's state. Hence, the state space in any such practical application would be very large and very high dimensional. The same holds for the action spaces. Thus, working with large, multidimensional state- and action-spaces should be considered the normal situation, while the examples presented in this section with their one-dimensional, small state spaces should be viewed as the exceptions.

1.3   VALUE FUNCTIONS

The obvious way of finding an optimal behavior in some MDP is to list all behaviors and then identify the ones that give the highest possible value for each initial state. Since, in general, there are too many behaviors, this plan is not viable. A better approach is based on computing value functions. In this approach, one first computes the so-called optimal value function, which then allows one to determine an optimal behavior with relative easiness.

The *optimal value*, $V^*(x)$, of state $x \in \mathcal{X}$ gives the highest achievable expected return when the process is started from state x. The function $V^* : \mathcal{X} \to \mathbb{R}$ is called the *optimal value function*. A behavior that achieves the optimal values in *all* states is *optimal*.

Deterministic stationary policies represent a special class of behaviors, which, as we shall see soon, play an important role in the theory of MDPs. They are specified by some mapping π, which maps states to actions (i.e., $\pi : \mathcal{X} \to \mathcal{A}$). Following π means that at any time $t \geq 0$ the action A_t is selected using

$$A_t = \pi(X_t). \tag{1.6}$$

More generally, a *stochastic stationary policy* (or just stationary policy) π maps states to distributions over the action space. When referring to such a policy π, we shall use $\pi(a|x)$ to denote the probability of action a being selected by π in state x. Note that if a stationary policy is followed in an MDP, i.e., if

$$A_t \sim \pi(\cdot \mid X_t), \qquad t \in \mathbb{N},$$

the state process $(X_t; t \geq 0)$ will be a (time-homogeneous) Markov chain. We will use Π_{stat} to denote the set of all stationary policies. For brevity, in what follows, we will often say just "policy" instead of "stationary policy", hoping that this will not cause confusion.

A stationary policy and an MDP induce what is called a *Markov reward processes* (MRP): An MRP is determined by the pair $\mathcal{M} = (\mathcal{X}, \mathcal{P}_0)$, where now $\mathcal{P}_0$ assigns a probability measure over $\mathcal{X} \times \mathbb{R}$ to each state. An MRP $\mathcal{M}$ gives rise to the stochastic process $((X_t, R_{t+1}); t \geq 0)$, where $(X_{t+1}, R_{t+1}) \sim \mathcal{P}_0(\cdot \mid X_t)$. (Note that $(Z_t; t \geq 0)$, $Z_t = (X_t, R_t)$ is a time-homogeneous Markov process, where R_0 is an arbitrary random variable, while $((X_t, R_{t+1}); t \geq 0)$ is a second-order Markov process.) Given a stationary policy π and the MDP $\mathcal{M} = (\mathcal{X}, \mathcal{A}, \mathcal{P}_0)$, the transition kernel of the MRP $(\mathcal{X}, \mathcal{P}_0^\pi)$ induced by π and $\mathcal{M}$ is defined using $\mathcal{P}_0^\pi(\cdot \mid x) = \sum_{a \in \mathcal{A}} \pi(a|x) \mathcal{P}_0(\cdot \mid x, a)$. An MRP is called finite if its state space is finite.

Let us now define value functions underlying stationary policies.[5] For this, let us fix some policy $\pi \in \Pi_{\text{stat}}$. The *value function*, $V^\pi : \mathcal{X} \to \mathbb{R}$, underlying π is defined by

$$V^\pi(x) = \mathbb{E}\left[\sum_{t=0}^\infty \gamma^t R_{t+1} \,\Big|\, X_0 = x\right], \quad x \in \mathcal{X}, \tag{1.7}$$

with the understanding *(i)* that the process $(R_t; t \geq 1)$ is the "reward-part" of the process $((X_t, A_t, R_{t+1}); t \geq 0)$ obtained when following policy π and *(ii)* X_0 is selected at random such that $\mathbb{P}(X_0 = x) > 0$ holds for all states x. This second condition makes the conditional expectation in (1.7) well-defined for every state. If the initial state distribution satisfies this condition, it has no influence on the definition of values.

The value function underlying an MRP is defined the same way and is denoted by V:

$$V(x) = \mathbb{E}\left[\sum_{t=0}^\infty \gamma^t R_{t+1} \,\Big|\, X_0 = x\right], \quad x \in \mathcal{X}.$$

It will also be useful to define the *action-value function*, $Q^\pi : \mathcal{X} \times \mathcal{A} \to \mathbb{R}$, underlying a policy $\pi \in \Pi_{\text{stat}}$ in an MDP: Assume that the first action A_0 is selected randomly such that $\mathbb{P}(A_0 = a) > 0$ holds for all $a \in \mathcal{A}$, while for the subsequent stages of the decision process the actions are chosen by following policy π. Let $((X_t, A_t, R_{t+1}); t \geq 0)$ be the resulting stochastic process, where X_0 is as in the definition of V^π. Then

$$Q^\pi(x, a) = \mathbb{E}\left[\sum_{t=0}^\infty \gamma^t R_{t+1} \,\Big|\, X_0 = x, A_0 = a\right], \quad x \in \mathcal{X}, a \in \mathcal{A}.$$

Similarly to $V^*(x)$, the *optimal action-value* $Q^*(x, a)$ at the state-action pair (x, a) is defined as the maximum of the expected return under the constraints that the process starts at state x, and the first action chosen is a. The underlying function $Q^* : \mathcal{X} \times \mathcal{A} \to \mathbb{R}$ is called the *optimal action-value function*.

The optimal value- and action-value functions are connected by the following equations:

$$
\begin{aligned}
V^*(x) &= \sup_{a \in \mathcal{A}} Q^*(x, a), & x \in \mathcal{X}, \\
Q^*(x, a) &= r(x, a) + \gamma \sum_{y \in \mathcal{X}} \mathcal{P}(x, a, y) V^*(y), & x \in \mathcal{X}, a \in \mathcal{A}.
\end{aligned}
$$

In the class of MDPs considered here, an optimal stationary policy always exists:

$$V^*(x) = \sup_{\pi \in \Pi_{\text{stat}}} V^\pi(x), \quad x \in \mathcal{X}.$$

[5]Value functions can also be defined underlying any behavior analogously to the definition given below.

In fact, any policy $\pi \in \Pi_{\text{stat}}$ which satisfies

$$\sum_{a \in \mathcal{A}} \pi(a|x) \, Q^*(x, a) = V^*(x) \tag{1.8}$$

simultaneously for all states $x \in \mathcal{X}$ is optimal. Notice that in order (1.8) to hold, $\pi(\cdot|x)$ must be concentrated on the set of actions that maximize $Q^*(x, \cdot)$. In general, given some action-value function, $Q : \mathcal{X} \times \mathcal{A} \to \mathbb{R}$, an action that maximizes $Q(x, \cdot)$ for some state x is called *greedy* with respect to Q in state x. A policy that chooses greedy actions only with respect to Q in *all* states is called *greedy w.r.t. Q*.

Thus, a greedy policy with respect to Q^* is optimal, i.e., the knowledge of Q^* alone is sufficient for finding an optimal policy. Similarly, knowing V^*, r and $\mathcal{P}$ also suffices to act optimally.

The next question is how to find V^* or Q^*. Let us start with the simpler question of how to find the value function of a policy:

Fact 1.3 Bellman Equations for Deterministic Policies   Fix an MDP $\mathcal{M} = (\mathcal{X}, \mathcal{A}, \mathcal{P}_0)$, a discount factor γ and deterministic policy $\pi \in \Pi_{\text{stat}}$. Let r be the immediate reward function of $\mathcal{M}$. Then V^π satisfies

$$V^\pi(x) = r(x, \pi(x)) + \gamma \sum_{y \in \mathcal{X}} \mathcal{P}(x, \pi(x), y) V^\pi(y), \qquad x \in \mathcal{X}. \tag{1.9}$$

This system of equations is called the *Bellman equation* for V^π. Define the *Bellman operator* underlying π, $T^\pi : \mathbb{R}^{\mathcal{X}} \to \mathbb{R}^{\mathcal{X}}$, by

$$(T^\pi V)(x) = r(x, \pi(x)) + \gamma \sum_{y \in \mathcal{X}} \mathcal{P}(x, \pi(x), y) V(y), \quad x \in \mathcal{X}.$$

With the help of T^π, Equation (1.9) can be written in the compact form

$$T^\pi V^\pi = V^\pi. \tag{1.10}$$

Note that this is a linear system of equations in V^π and T^π is an affine linear operator. If $0 < \gamma < 1$ then T^π is a maximum-norm contraction and the fixed-point equation $T^\pi V = V$ has a unique solution. When the state space $\mathcal{X}$ is finite, say, it has D states, $\mathbb{R}^{\mathcal{X}}$ can be identified with the D-dimensional Euclidean space and $V \in \mathbb{R}^{\mathcal{X}}$ can be thought of as a D-dimensional vector: $V \in \mathbb{R}^D$. With this identification, $T^\pi V$ can also be written as $r^\pi + \gamma P^\pi V$ with an appropriately defined vector $r^\pi \in \mathbb{R}^D$ and matrix $P^\pi \in \mathbb{R}^{D \times D}$. In this case, (1.10) can be written in the form

$$r^\pi + \gamma P^\pi V^\pi = V^\pi. \tag{1.11}$$

The above facts also hold true in MRPs, where the Bellman operator $T : \mathbb{R}^{\mathcal{X}} \to \mathbb{R}^{\mathcal{X}}$ is defined by

$$(TV)(x) = r(x) + \gamma \sum_{y \in \mathcal{X}} \mathcal{P}(x, y) V(y), \quad x \in \mathcal{X}.$$

The optimal value function is known to satisfy a certain fixed-point equation:

Fact 1.4   Bellman Optimality Equations   The optimal value function satisfies the fixed-point equation

$$V^*(x) = \sup_{a \in \mathcal{A}} \left\{ r(x, a) + \gamma \sum_{y \in \mathcal{X}} \mathcal{P}(x, a, y) V^*(y) \right\}, \qquad x \in \mathcal{X}. \tag{1.12}$$

Define the *Bellman optimality operator* operator, $T^* : \mathbb{R}^{\mathcal{X}} \to \mathbb{R}^{\mathcal{X}}$, by

$$(T^* V)(x) = \sup_{a \in \mathcal{A}} \left\{ r(x, a) + \gamma \sum_{y \in \mathcal{X}} \mathcal{P}(x, a, y) V(y) \right\}, \qquad x \in \mathcal{X}. \tag{1.13}$$

Note that this is a nonlinear operator due to the presence of sup. With the help of T^*, Equation (1.12) can be written compactly as

$$T^* V^* = V^*.$$

If $0 < \gamma < 1$, then T^* is a maximum-norm contraction, and the fixed-point equation $T^* V = V$ has a unique solution. In order to minimize clutter, in what follows we will write expressions like

$(T^\pi V)(x)$ as $T^\pi V(x)$, with the understanding that the application of operator T^π takes precedence to the application of the point evaluation operator, "$\cdot (x)$".

The action-value functions underlying a policy (or an MRP) and the optimal action-value function also satisfy some fixed-point equations similar to the previous ones:

Fact 1.5   Bellman Operators and Fixed-point Equations for Action-value Functions   With a slight abuse of notation, define $T^\pi : \mathbb{R}^{\mathcal{X} \times \mathcal{A}} \to \mathbb{R}^{\mathcal{X} \times \mathcal{A}}$ and $T^* : \mathbb{R}^{\mathcal{X} \times \mathcal{A}} \to \mathbb{R}^{\mathcal{X} \times \mathcal{A}}$ as follows:

$$T^\pi Q(x, a) \;=\; r(x, a) + \gamma \sum_{y \in \mathcal{X}} \mathcal{P}(x, a, y) Q(y, \pi(x)), \qquad (x, a) \in \mathcal{X} \times \mathcal{A}, \tag{1.14}$$

$$T^* Q(x, a) \;=\; r(x, a) + \gamma \sum_{y \in \mathcal{X}} \mathcal{P}(x, a, y) \sup_{a' \in \mathcal{A}} Q(y, a'), \qquad (x, a) \in \mathcal{X} \times \mathcal{A}. \tag{1.15}$$

Note that T^π is again affine linear, while T^* is nonlinear. The operators T^π and T^* are maximum-norm contractions. Further, the action-value function of π, Q^π, satisfies $T^\pi Q^\pi = Q^\pi$ and Q^π is the unique solution to this fixed-point equation. Similarly, the optimal action-value function, Q^*, satisfies $T^* Q^* = Q^*$ and Q^* is the unique solution to this fixed-point equation.

1.4   DYNAMIC PROGRAMMING ALGORITHMS FOR SOLVING MDPS

The above facts provide the basis for the *value-* and *policy-iteration* algorithms.

Value iteration generates a sequence of value functions

$$V_{k+1} = T^* V_k, \quad k \geq 0,$$

where V_0 is arbitrary. Thanks to Banach's fixed-point theorem, $(V_k; k \geq 0)$ converges to V^* at a geometric rate.

Value iteration can also be used in conjunction with action-value functions; in which case, it takes the form

$$Q_{k+1} = T^* Q_k, \quad k \geq 0,$$

which again converges to Q^* at a geometric rate. The idea is that once V_k (or Q_k) is close to V^* (resp., Q^*), a policy that is greedy with respect to V_k (resps., Q_k) will be close-to-optimal.

In particular, the following bound is known to hold: Fix an action-value function Q and let π be a greedy policy w.r.t. Q. Then the value of policy π can be lower bounded as follows (e.g., Singh and Yee, 1994, Corollary 2):

$$V^\pi(x) \geq V^*(x) - \frac{2}{1-\gamma} \|Q - Q^*\|_\infty, \quad x \in \mathcal{X}. \tag{1.16}$$

Policy iteration works as follows. Fix an arbitrary initial policy π_0. At iteration $k > 0$, compute the action-value function underlying π_k (this is called the policy evaluation step). Next, given Q^{π_k}, define π_{k+1} as a policy that is greedy with respect to Q^{π_k} (this is called the policy improvement step). After k iterations, policy iteration gives a policy not worse than the policy that is greedy w.r.t. to the value function computed using k iterations of value iteration if the two procedures are started with the same initial value function. However, the computational cost of a single step in policy iteration is much higher (because of the policy evaluation step) than that of one update in value iteration.

CHAPTER 2

Value Prediction Problems

In this section, we consider the problem of estimating the value function V underlying some Markov reward process (MRP). Value prediction problems arise in a number of ways: Estimating the probability of some future event, the expected time until some event occurs, or the (action-)value function underlying some policy in an MDP are all value prediction problems. Specific applications are estimating the failure probability of a large power grid (Frank et al., 2008) or estimating taxi-out times of flights on busy airports (Balakrishna et al., 2008), just to mention two of the many possibilities.

Since the value of a state is defined as the expectation of the random return when the process is started from the given state, an obvious way of estimating this value is to compute an average over multiple independent realizations started from the given state. This is an instance of the so-called *Monte-Carlo method*. Unfortunately, the variance of the returns can be high, which means that the quality of the estimates will be poor. Also, when interacting with a system in a closed-loop fashion (i.e., when estimation happens while interacting with the system), it might be impossible to reset the state of the system to some particular state. In this case, the Monte-Carlo technique cannot be applied without introducing some additional bias. *Temporal difference (TD) learning* (Sutton, 1984, 1988), which is without doubt one of the most significant ideas in reinforcement learning, is a method that can be used to address these issues.

2.1 TEMPORAL DIFFERENCE LEARNING IN FINITE STATE SPACES

The unique feature of TD learning is that it uses *bootstrapping*: predictions are used as targets during the course of learning. In this section, we first introduce the most basic TD algorithm and explain how bootstrapping works. Next, we compare TD learning to (vanilla) Monte-Carlo methods, we and argue that both of them have their own merits. Finally, we present the TD(λ) algorithm that unifies the two approaches. Here we consider only the case of small, finite MRPs, when the value-estimates of all the states can be stored in the main memory of a computer in an array or table, which is known as the *tabular case* in the reinforcement learning literature. Extensions of the ideas presented here to large state spaces, when a tabular representations is not feasible, will be described in the subsequent sections.

2.1.1 TABULAR TD(0)

Fix some finite Markov Reward Process $\mathcal{M}$. We wish to estimate the value function V underlying $\mathcal{M}$ given a realization $((X_t, R_{t+1}); t \geq 0)$ of $\mathcal{M}$. Let $\hat{V}_t(x)$ denote the estimate of state x at time t

---

Algorithm 1 The function implementing the tabular TD(0) algorithm. This function must be called after each transition.

---

function TD0(X, R, Y, V)
Input: X is the last state, Y is the next state, R is the immediate reward associated with this transition, V is the array storing the current value estimates
1: $\delta \leftarrow R + \gamma \cdot V[Y] - V[X]$
2: $V[X] \leftarrow V[X] + \alpha \cdot \delta$
3: **return** V

(say, $\hat{V}_0 \equiv 0$). In the tth step TD(0) performs the following calculations:

$$\delta_{t+1} = R_{t+1} + \gamma \hat{V}_t(X_{t+1}) - \hat{V}_t(X_t),$$
$$\hat{V}_{t+1}(x) = \hat{V}_t(x) + \alpha_t \, \delta_{t+1} \, \mathbb{I}_{\{X_t=x\}}, \tag{2.1}$$
$$x \in \mathcal{X}.$$

Here the *step-size* sequence $(\alpha_t; t \geq 0)$ consists of (small) nonnegative numbers chosen by the user. Algorithm 1 shows the pseudocode of this algorithm.

A closer inspection of the update equation reveals that the only value changed is the one associated with X_t, i.e., the state just visited (cf. line 2 of the pseudocode). Further, when $\alpha_t \leq 1$, the value of X_t is moved towards the "target" $R_{t+1} + \gamma \hat{V}_t(X_{t+1})$. Since the target depends on the estimated value function, the algorithm uses *bootstrapping*. The term "temporal difference" in the name of the algorithm comes from that δ_{t+1} is defined as the difference between values of states corresponding to successive time steps. In particular, δ_{t+1} is called a *temporal difference error*.

Just like many other algorithms in reinforcement learning, tabular TD(0) is a stochastic approximation (SA) algorithm. It is easy to see that if it converges, then it must converge to a function $\hat{V}$ such that the expected temporal difference given $\hat{V}$,

$$F\hat{V}(x) \stackrel{\text{def}}{=} \mathbb{E}\left[R_{t+1} + \gamma \hat{V}(X_{t+1}) - \hat{V}(X_t) \,\Big|\, X_t = x \right],$$

is zero for all states x, at least for all states that are sampled infinitely often. A simple calculation shows that $F\hat{V} = T\hat{V} - \hat{V}$, where T is the Bellman-operator underlying the MRP considered. By Fact 1.3, $F\hat{V} = 0$ has a unique solution, the value function V. Thus, if TD(0) converges (and all states are sampled infinitely often) then it must converge to V.

To study the algorithm's convergence properties, for simplicity, assume that $(X_t; t \in \mathbb{N})$ is a stationary, ergodic Markov chain.[1] Further, identify the approximate value functions $\hat{V}_t$ with D-dimensional vectors as before (e.g., $\hat{V}_{t,i} = \hat{V}_t(x_i)$, $i = 1, \ldots, D$, where $D = |\mathcal{X}|$ and $\mathcal{X} = \{x_1, \ldots, x_D\}$). Then, assuming that the step-size sequence satisfies the *Robbins-Monro (RM)*

---

[1] Remember that a Markov chain $(X_t; t \in \mathbb{N})$ is ergodic it is irreducible, aperiodic and positive recurrent. Practically, this means that the law of large number holds for sufficiently regular functions of the chain.

conditions,

$$\sum_{t=0}^{\infty} \alpha_t = \infty, \qquad \sum_{t=0}^{\infty} \alpha_t^2 < +\infty,$$

the sequence $(\hat{V}_t \in \mathbb{R}^D; t \in \mathbb{N})$ will track the trajectories of the ordinary differential equation (ODE)

$$\dot{v}(t) = c \, F(v(t)), \quad t \geq 0, \tag{2.2}$$

where $c = 1/D$ and $v(t) \in \mathbb{R}^D$ (e.g., Borkar, 1998). Borrowing the notation used in (1.11), the above ODE can be written as

$$\dot{v} = r + (\gamma P - I)v.$$

Note that this is a *linear* ODE. Since the eigenvalues of $\gamma P - I$ all lie in the open left half complex plane, this ODE is globally asymptotically stable. From this, using standard results of SA it follows that $\hat{V}_t$ converges almost surely to V.

On step-sizes   Since many of the algorithms that we will discuss use step-sizes, it is worthwhile spending some time on discussing their choice. A simple step-size sequence that satisfies the above conditions is $\alpha_t = c/t$, with $c > 0$. More generally, any step-size sequence of the form $\alpha_t = ct^{-\eta}$ will work as long as $1/2 < \eta \leq 1$. Of these step-size sequences, $\eta = 1$ gives the smallest step-sizes. Asymptotically, this choice will be the best, but from the point of view of the transient behavior of the algorithm, choosing η closer to $1/2$ will work better (since with this choice the step-sizes are bigger and thus the algorithm will make larger moves). It is possible to do even better than this. In fact, a simple method, called iterate-averaging due to Polyak and Juditsky (1992), is known to achieve the best possible asymptotic rate of convergence. However, despite its appealing theoretical properties, iterate-averaging is rarely used in practice. In fact, in practice people often use constant step-sizes, which clearly violates the RM conditions. This choice is justified based on two grounds: First, the algorithms are often used in a non-stationary environment (i.e., the policy to be evaluated might change). Second, the algorithms are often used only in the small sample regime. (When a constant step-size is used, the parameters converge in distribution. The variance of the limiting distribution will be proportional to the step-size chosen.) There is also a great deal of work going into developing methods that tune step-sizes automatically, see (Sutton, 1992; Schraudolph, 1999; George and Powell, 2006) and the references therein. However, the jury is still out on which of these methods is the best.

    With a small change, the algorithm can also be used on an observation sequence of the form $((X_t, R_{t+1}, Y_{t+1}); t \geq 0)$, where $(X_t; t \geq 0)$ is an *arbitrary* ergodic Markov chain over $\mathcal{X}$, $(Y_{t+1}, R_{t+1}) \sim \mathcal{P}_0(\cdot \mid X_t)$. The change concerns the definition of temporal differences:

$$\delta_{t+1} = R_{t+1} + \gamma \hat{V}(Y_{t+1}) - \hat{V}(X_t).$$

Then, with no extra conditions, $\hat{V}_t$ still converges almost surely to the value function underlying the MRP $(\mathcal{X}, \mathcal{P}_0)$. In particular, the distribution of the states $(X_t; t \geq 0)$ does not play a role here.

This is interesting for multiple reasons. For example, if the samples are generated using a simulator, we may be able to control the distribution of the states $(X_t; t \geq 0)$ independently of the MRP. This might be useful to counterbalance any unevenness in the stationary distribution underlying the Markov kernel $\mathcal{P}$. Another use is to learn about some *target policy* in an MDP while following some other policy, often called the *behavior* policy. Assume for simplicity that the target policy is deterministic. Then $((X_t, R_{t+1}, Y_{t+1}), t \geq 0)$ could be obtained by skipping all those state-action-reward-next state quadruples in the trajectory generated by using the behavior policy, where the action taken does not match the action that would have been taken in the given state by the target policy, while keeping the rest. This technique might allow one to learn about multiple policies at the same time (more generally, about multiple long-term prediction problems). When learning about one policy, while following another is called *off-policy learning*. Because of this, we shall also call learning based on triplets $((X_t, R_{t+1}, Y_{t+1}); t \geq 0)$ when $Y_{t+1} \neq X_{t+1}$ off-policy learning. A third, technical use is when the goal is to apply the algorithm to an episodic problem. In this case, the triplets (X_t, R_{t+1}, Y_{t+1}) are chosen as follows: First, Y_{t+1} is sampled from the transition kernel $\mathcal{P}(X, \cdot)$. If Y_{t+1} is not a terminal state, we let $X_{t+1} = Y_{t+1}$; otherwise, $X_{t+1} \sim \mathcal{P}_0(\cdot)$, where $\mathcal{P}_0$ is a user-chosen distribution over $\mathcal{X}$. In other words, when a terminal state is reached, the process is restarted from the initial state distribution $\mathcal{P}_0$. The period between the time of a restart from $\mathcal{P}_0$ and reaching a terminal state is called an *episode* (hence the name of episodic problems). This way of generating a sample shall be called *continual sampling with restarts from* $\mathcal{P}_0$.

Being a standard linear SA method, the rate of convergence of tabular TD(0) will be of the usual order $O(1/\sqrt{t})$ (consult the paper by Tadić (2004) and the references therein for precise results). However, the constant factor in the rate will be largely influenced by the choice of the step-size sequence, the properties of the kernel $\mathcal{P}_0$ and the value of γ.

2.1.2   EVERY-VISIT MONTE-CARLO

As mentioned before, one can also estimate the value of a state by computing sample means, giving rise to the so-called *every visit Monte-Carlo method*. Here we define more precisely what we mean by this and compare the resulting method to TD(0).

To firm up the ideas, consider some episodic problem (otherwise, it is impossible to finitely compute the return of a given state since the trajectories are infinitely long). Let the underlying MRP be $\mathcal{M} = (\mathcal{X}, \mathcal{P}_0)$ and let $((X_t, R_{t+1}, Y_{t+1}); t \geq 0)$ be generated by continual sampling in $\mathcal{M}$ with restarts from some distribution $\mathcal{P}_0$ defined over $\mathcal{X}$. Let $(T_k; k \geq 0)$ be the sequence of times when an episode starts (thus, for each k, X_{T_k} is sampled from $\mathcal{P}_0$). For a given time t, let $k(t)$ be the unique episode index such that $t \in [T_k, T_{k+1})$. Let

$$\mathcal{R}_t = \sum_{s=t}^{T_{k(t)+1}-1} \gamma^{s-t} R_{s+1} \tag{2.3}$$

Algorithm 2 The function that implements the every-visit Monte-Carlo algorithm to estimate value functions in episodic MDPs. This routine must be called at the end of each episode with the state-reward sequence collected during the episode. Note that the algorithm as shown here has linear time- and space-complexity in the length of the episodes.

---

function EVERYVISITMC($X_0, R_1, X_1, R_2, \ldots, X_{T-1}, R_T, V$)

Input: X_t is the state at time t, R_{t+1} is the reward associated with the tth transition, T is the length of the episode, V is the array storing the current value function estimate

1: $sum \leftarrow 0$
2: **for** $t \leftarrow T - 1$ **downto** 0 **do**
3:      $sum \leftarrow R_{t+1} + \gamma \cdot sum$
4:      $target[X_t] \leftarrow sum$
5:      $V[X_t] \leftarrow V[X_t] + \alpha \cdot (target[X_t] - V[X_t])$
6: **end for**
7: **return** V

---

denote the return from time t on until the end of the episode. Clearly, $V(x) = \mathbb{E}\left[\mathcal{R}_t | X_t = x\right]$, for any state x such that $\mathbb{P}(X_t = x) > 0$. Hence, a sensible way of updating the estimates is to use

$$\hat{V}_{t+1}(x) = \hat{V}_t(x) + \alpha_t(\mathcal{R}_t - \hat{V}_t(x))\, \mathbb{I}_{\{X_t = x\}}, \qquad x \in \mathcal{X}.$$

Monte-Carlo methods such as the above one, since they use multi-step predictions of the return (cf. Equation (2.3)), are called *multi-step* methods. The pseudo-code of this update-rule is shown as Algorithm 2.

This algorithm is again an instance of stochastic approximation. As such, its behavior is governed by the ODE $\dot{v}(t) = V - v(t)$. Since the unique globally asymptotically stable equilibrium of this ODE is V, $\hat{V}_t$ again converges to V almost surely. Since both algorithms achieve the same goal, one may wonder which algorithm is better.

TD(0) or Monte-Carlo?   First, let us consider an example when TD(0) converges faster. Consider the undiscounted episodic MRP shown on Figure 2.1. The initial states are either 1 or 2. With high probability the process starts at state 1, while the process starts at state 2 less frequently. Consider now how TD(0) will behave at state 2. By the time state 2 is visited the kth time, on the average state 3 has already been visited $10\,k$ times. Assume that $\alpha_t = 1/(t + 1)$. At state 3 the TD(0) update reduces to averaging the Bernoulli rewards incurred upon leaving state 3. At the kth visit of state 2, $\text{Var}\left[\hat{V}_t(3)\right] \approx 1/(10\,k)$ (clearly, $\mathbb{E}\left[\hat{V}_t(3)\right] = V(3) = 0.5$). Thus, the target of the update of state 2 will be an estimate of the true value of state 2 with accuracy increasing with k. Now, consider the Monte-Carlo method. The Monte-Carlo method ignores the estimate of the value of state 3 and uses the Bernoulli rewards directly. In particular, $\text{Var}\left[\mathcal{R}_t | X_t = 2\right] = 0.25$, i.e., the variance of the target does not change with time. On this example, this makes the Monte-Carlo method slower to converge, showing that sometimes bootstrapping might indeed help.

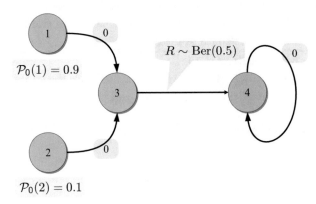

Figure 2.1: An episodic Markov reward process. In this example, all transitions are deterministic. The reward is zero, except when transitioning from state 3 to state 4, when it is given by a Bernoulli random variable with parameter 0.5. State 4 is a terminal state. When the process reaches the terminal state, it is reset to start at state 1 or 2. The probability of starting at state 1 is 0.9, while the probability of starting at state 2 is 0.1.

To see an example when bootstrapping is not helpful, imagine that the problem is modified so that the reward associated with the transition from state 3 to state 4 is made deterministically equal to one. In this case, the Monte-Carlo method becomes faster since $\mathcal{R}_t = 1$ is the true target value, while for the value of state 2 to get close to its true value, TD(0) has to wait until the estimate of the value at state 3 becomes close to its true value. This slows down the convergence of TD(0). In fact, one can imagine a longer chain of states, where state $i + 1$ follows state i, for $i \in \{1, \ldots, N\}$ and the only time a nonzero reward is incurred is when transitioning from state $N - 1$ to state N. In this example, the rate of convergence of the Monte-Carlo method is not impacted by the value of N, while TD(0) would get slower with N increasing (for an informal argument, see Sutton, 1988; for a formal one with exact rates, see Beleznay et al., 1999).

2.1.3   TD(λ): UNIFYING MONTE-CARLO AND TD(0)

The previous examples show that both Monte-Carlo and TD(0) have their own merits. Interestingly, there is a way to unify these approaches. This is achieved by the so-called TD(λ) family of methods (Sutton, 1984, 1988). Here, $\lambda \in [0, 1]$ is a parameter that allows one to interpolate between the Monte-Carlo and TD(0) updates: $\lambda = 0$ gives TD(0) (hence the name of TD(0)), while $\lambda = 1$, i.e., TD(1) is equivalent to a Monte-Carlo method. In essence, given some $\lambda > 0$, the targets in the

TD(λ) update are given as some mixture of the multi-step return predictions

$$\mathcal{R}_{t:k} = \sum_{s=t}^{t+k} \gamma^{s-t} R_{s+1} + \gamma^{k+1} \hat{V}_t(X_{t+k+1}),$$

where the mixing coefficients are the exponential weights $(1 - \lambda)\lambda^k$, $k \geq 0$. Thus, for $\lambda > 0$ TD(λ) will be a multi-step method. The algorithm is made incremental by the introduction of the so-called eligibility traces.

In fact, the eligibility traces can be defined in multiple ways and hence TD(λ) exists in correspondingly many multiple forms. The update rule of TD(λ) with the so-called *accumulating traces* is as follows:

$$\begin{aligned}
\delta_{t+1} &= R_{t+1} + \gamma \hat{V}_t(X_{t+1}) - \hat{V}_t(X_t), \\
z_{t+1}(x) &= \mathbb{I}_{\{x=X_t\}} + \gamma\lambda\, z_t(x), \\
\hat{V}_{t+1}(x) &= \hat{V}_t(x) + \alpha_t\, \delta_{t+1}\, z_{t+1}(x), \\
z_0(x) &= 0, \\
& \quad x \in \mathcal{X}.
\end{aligned}$$

Here $z_t(x)$ is the *eligibility trace* of state x. The rationale of the name is that the value of $z_t(x)$ modulates the influence of the TD error on the update of the value stored at state x. In another variant of the algorithm, the eligibility traces are updated according to

$$z_{t+1}(x) = \max(\mathbb{I}_{\{x=X_t\}}, \gamma\lambda\, z_t(x)), \qquad x \in \mathcal{X}.$$

This is called the *replacing traces* update. In these updates, the *trace-decay parameter* λ controls the amount of bootstrapping: When $\lambda = 0$ the above algorithms become identical to TD(0) (since $\lim_{\lambda \to 0+} (1 - \lambda) \sum_{k \geq 0} \lambda^k \mathcal{R}_{t:k} = \mathcal{R}_{t:0} = R_{t+1} + \gamma \hat{V}_t(X_{t+1})$). When $\lambda = 1$, we get the TD(1) algorithm, which with accumulating traces will simulate the previously described every-visit Monte-Carlo algorithm in episodic problems. (For an exact equivalence, one needs to assume that the value updates happen only at the end of trajectories, up to which point the updates are just accumulated. The statement then follows because the discounted sum of temporal differences along a trajectory from a start state to a terminal state telescopes and gives the sum of rewards along the trajectory.) Replacing traces and $\lambda = 1$ correspond to a version of the Monte-Carlo algorithm where a state is updated only when it is encountered for the first time in a trajectory. The corresponding algorithm is called *first-visit Monte-Carlo method*. The formal correspondence between the first-visit Monte-Carlo method and TD(1) with replacing traces is known to hold for the undiscounted case only (Singh and Sutton, 1996). Algorithm 3 gives the pseudocode corresponding to the variant with replacing traces.

In practice, the best value of λ is determined by trial and error. In fact, the value of λ can be changed even during the algorithm, without impacting convergence. This holds for a wide range of other possible eligibility trace updates (for precise conditions, see Bertsekas and Tsitsiklis, 1996,

---

Algorithm 3 The function that implements the tabular TD(λ) algorithm with replacing traces. This function must be called after each transition.

---

function TDLAMBDA(X, R, Y, V, z)

Input: X is the last state, Y is the next state, R is the immediate reward associated with this transition, V is the array storing the current value function estimate, z is the array storing the eligibility traces

1: $\delta \leftarrow R + \gamma \cdot V[Y] - V[X]$
2: **for all** $x \in \mathcal{X}$ **do**
3:     $z[x] \leftarrow \gamma \cdot \lambda \cdot z[x]$
4:     **if** $X = x$ **then**
5:         $z[x] \leftarrow 1$
6:     **end if**
7:     $V[x] \leftarrow V[x] + \alpha \cdot \delta \cdot z[x]$
8: **end for**
9: **return** (V, z)

Section 5.3.3 and 5.3.6). The replacing traces version of the algorithm is believed to perform better in practice (for some examples when this happens, consult Sutton and Barto, 1998, Section 7.8). It has been noted that $\lambda > 0$ is helpful when the learner has only partial knowledge of the state, or (in the related situation) when function approximation is used to approximate the value functions in a large state space – the topic of the next section.

In summary, TD(λ) allows one to estimate value functions in MRPs. It generalizes Monte-Carlo methods, it can be used in non-episodic problems, and it allows for bootstrapping. Further, by appropriately tuning λ it can converge significantly faster than Monte-Carlo methods or TD(0).

2.2 ALGORITHMS FOR LARGE STATE SPACES

When the state space is large (or infinite), it is not feasible to keep a separate value for each state in the memory. In such cases, we often seek an estimate of the values in the form

$$V_\theta(x) = \theta^\top \varphi(x), \qquad x \in \mathcal{X},$$

where $\theta \in \mathbb{R}^d$ is a vector of parameters and $\varphi : \mathcal{X} \to \mathbb{R}^d$ is a mapping of states to d-dimensional vectors. For state x, the components $\varphi_i(x)$ of the vector $\varphi(x)$ are called the *features* of state x and φ is called a *feature extraction* method. The individual functions $\varphi_i : \mathcal{X} \to \mathbb{R}$ defining the components of φ are called *basis functions*.

Examples of function approximation methods Given access to the state, the features (or basis functions) can be constructed in a great many different ways. If $x \in \mathbb{R}$ (i.e., $\mathcal{X} \subset \mathbb{R}$) one may use a polynomial, Fourier, or wavelet basis up to some order. For example, in the case of a polynomial basis, $\varphi(x) = (1, x, x^2, \dots, x^{d-1})^\top$, or, an orthogonal system of polynomials if a suitable measure (such as the stationary distribution) over the states is available. This latter choice may help to increase the convergence speed of the incremental algorithms that we will discuss soon.

In the case of multidimensional state spaces, the *tensor product construction* is a commonly used way to construct features given features of the states' individual components. The tensor product construction works as follows: Imagine that $\mathcal{X} \subset \mathcal{X}_1 \times \mathcal{X}_2 \times \ldots \times \mathcal{X}_k$. Let $\varphi^{(i)} : \mathcal{X}_i \to \mathbb{R}^{d_i}$ be a feature extractor defined for the ith state component. The tensor product $\varphi = \varphi^{(1)} \otimes \ldots \otimes \varphi^{(k)}$ feature extractor will have $d = d_1 d_2 \ldots d_k$ components, which can be conveniently indexed using multi-indices of the form $(i_1, \ldots, i_k)$, $1 \leq i_j \leq d_j$, $j = 1, \ldots, k$. Then $\varphi_{(i_1, \ldots, i_k)}(x) = \varphi_{i_1}^{(1)}(x_1) \varphi_{i_2}^{(2)}(x_2) \ldots \varphi_{i_k}^{(k)}(x_k)$. When $\mathcal{X} \subset \mathbb{R}^k$, one particularly popular choice is to use radial basis function (RBF) networks, when $\varphi^{(i)}(x_i) = (G(|x_i - x_i^{(1)}|), \ldots, G(|x_i - x_i^{(d_i)}|))^\top$. Here $x_i^{(j)} \in \mathbb{R}$ $(j = 1, \ldots, d_i)$ is fixed by the user and G is a suitable function. A typical choice for G is $G(z) = \exp(-\eta z^2)$ where $\eta > 0$ is a scale parameter. The tensor product construct in this cases places Gaussians at points of a regular grid and the ith basis function becomes

$$\varphi_i(x) = \exp(-\eta \|x - x^{(i)}\|^2),$$

where $x^{(i)} \in \mathcal{X}$ now denotes a point on a regular $d_1 \times \ldots \times d_k$ grid. A related method is to use *kernel smoothing*:

$$V_\theta(x) \;\; = \;\; \frac{\sum_{i=1}^d \theta_i \, G(\|x - x^{(i)}\|)}{\sum_{j=1}^d G(\|x - x^{(j)}\|)} = \sum_{i=1}^d \theta_i \, \frac{G(\|x - x^{(i)}\|)}{\sum_{j=1}^d G(\|x - x^{(j)}\|)}. \tag{2.4}$$

More generally, one may use $V_\theta(x) = \sum_{i=1}^d \theta_i s_i(x)$, where $s_i \geq 0$ and $\sum_{i=1}^d s_i(x) \equiv 1$ holds for any $x \in \mathcal{X}$. In this case, we say that V_θ is an *averager*. Averagers are important in reinforcement learning because the mapping $\theta \mapsto V_\theta$ is a non-expansion in the max-norm, which makes them "well-behaved" when used together with approximate dynamic programming.

An alternative to the above is to use binary features, i.e., when $\varphi(x) \in \{0, 1\}^d$. Binary features may be advantageous from a computational point of view: when $\varphi(x) \in \{0, 1\}^d$ then $V_\theta(x) = \sum_{i:\varphi_i(x)=1} \theta_i$. Thus, the value of state x can be computed at the cost of s additions if $\varphi(x)$ is s-*sparse* (i.e., if only s elements of $\varphi(x)$ are non-zero), provided that there is a direct way of computing the index of the non-zero components of the feature vector.

This is the case when the *state aggregation* is used to define the features. In this case, the coordinate functions of φ (the individual features) correspond to indicators of non-overlapping regions of the state space $\mathcal{X}$ whose union covers $\mathcal{X}$ (i.e., the regions form a partition of the state space). Clearly, in this case, $\theta^\top \varphi(x)$ will be constant over the individual regions, thus state aggregation essentially "discretizes" the state space. A state aggregator function approximator is also an averager.

Another choice that leads to binary features is *tile coding* (originally called CMAC, Albus, 1971, 1981). In the simplest version of tile coding, the basis functions of φ correspond to indicator functions of multiple shifted partitions (tilings) of the state space: if s tilings are used, φ will be s-sparse. To make tile coding an effective function approximation method, the offsets of the tilings corresponding to different dimensions should be different.

The curse of dimensionality   The issue with tensor product constructions, state aggregation and straightforward tile coding is that when the state space is high dimensional they quickly become intractable: For example, a tiling of $[0, 1]^D$ with cubical regions with side-lengths of ε gives rise to $d = \varepsilon^{-D}$-dimensional feature- and parameter-vectors. If $\varepsilon = 1/2$ and $D = 100$, we get the enormous number $d \approx 10^{30}$. This is problematic since state-representations with hundreds of dimensions are common in applications. At this stage, one may wonder if it is possible at all to successfully deal with applications when the state lives in a high dimensional space. What often comes at rescue is that the actual problem complexity might be much lower than what is predicted by merely counting the number of dimensions of the state variable (although, there is no guarantee that this happens). To see why sometimes this holds, note that the same problem can have multiple representations, some of which may come with low-dimensional state variables, some with high. Since, in many cases, the state-representation is chosen by the user in a conservative fashion, it may happen that in the chosen representation many of the state variables are irrelevant. It may also happen that the states that are actually encountered lie on (or lie close to) a low-dimensional submanifold of the chosen high dimensional "state-space".

To illustrate this, imagine an industrial robot arm with say 3 joints and 6 degrees of freedom. The intrinsic dimensionality of the state is then 12, twice the number of degrees of freedom of the arm since the dynamics is second-order. One (approximate) state representation is to take high resolution camera images of the arm in close succession (to account for the dynamics) from multiple angles (to account for occlusions). The dimensionality of the chosen state representation will easily be in the range of millions, yet the *intrinsic dimensionality* will still be 12. In fact, the more cameras we have, the higher the dimensionality will be. A simple-minded approach, which aims for minimizing the dimensionality would suggest to use as few cameras as possible. But more information should not hurt! Therefore, the quest should be for clever algorithms and function approximation methods that can deal with high-dimensional but low complexity problems.

Possibilities include using strip-like tilings combined with hash functions, interpolators that use low-discrepancy grids (Lemieux, 2009, Chapter 5 and 6), or random projections (Dasgupta and Freund, 2008). Nonlinear function approximation methods (examples of which include neural networks with sigmoidal transfer functions in the hidden layers or RBF networks where the centers are also considered as parameters) and nonparametric techniques also hold great promise.

Nonparametric methods   In a *nonparametric method*, the user does not start with a fixed finite-dimensional representation, such as in the previous examples, but allows for the representation to grow and change as needed. For example, in a *k-nearest neighbor method* for regression, given the data $\mathcal{D}_n = [(x_1, v_1), \dots, (x_n, v_n)]$, where $x_i \in \mathbb{R}^k$, $v_i \in \mathbb{R}$, the value at location x is predicted using

$$V_{\mathcal{D}}^{(k)}(x) = \sum_{i=1}^{n} v_i \, \frac{K_{\mathcal{D}}^{(k)}(x, x_i)}{k},$$

where $K_{\mathcal{D}}^{(k)}(x, x')$ is one when x' is closer to x then the kth closest neighbor of x in $\mathcal{D}$ and is zero otherwise. Note that $k = \sum_{j=1}^{n} K_{\mathcal{D}}^{(k)}(x, x_j)$. Replacing k in the above expression with this sum and replacing $K_{\mathcal{D}}^{(k)}(x, \cdot)$ with some other data based kernel $K_{\mathcal{D}}$ (e.g., a Gaussian centered around x with standard deviation proportional to the distance to the kth nearest neighbor), we arrive at *nonparametric kernel smoothing*:

$$V_{\mathcal{D}}^{(k)}(x) = \sum_{i=1}^{n} v_i \frac{K_{\mathcal{D}}(x, x_i)}{\sum_{j=1}^{n} K_{\mathcal{D}}(x, x_j)},$$

which should be compared to its parametric counterpart (2.4). Other examples include methods that work by finding an appropriate function in some large (infinite dimensional) function space that fits an empirical error. The function space is usually a Reproducing Kernel Hilbert space, which is a convenient choice from the point of view of optimization. In special cases, we get spline smoothers (Wahba, 2003) and Gaussian process regression (Rasmussen and Williams, 2005). Another idea is to split the input space recursively into finer regions using some heuristic criterion and then predict with some simple method the values in the leafs, leading to tree-based methods. The border between parametric and nonparametric methods is blurry. For example, a linear predictor when the number of basis functions is allowed to change (i.e., when new basis functions are introduced as needed) becomes a nonparametric method. Thus, when one experiments with different feature extraction methods, from the point of view of the overall tuning process, we can say that one really uses a nonparametric technique. In fact, if we take this viewpoint, it follows that in practice "true" parametric methods are rarely used if they are used at all.

The advantage of nonparametric methods is their inherent flexibility. However, this comes usually at the price of increased computational complexity. Therefore, when using nonparametric methods, efficient implementations are important (e.g., one should use k-D trees when implementing nearest neighbor methods, or the Fast Gaussian Transform when implementing a Gaussian smoother). Also, nonparametric methods must be carefully tuned as they can easily *overfit* or *underfit*. For example, in a k-nearest neighbor method if k is too large, the method is going to introduce too much smoothing (i.e., it will underfit), while if k is too small, it will fit to the noise (i.e., overfit). Overfitting will be further discussed in Section 2.2.4. For more information about nonparametric regression, the reader is advised to consult the books by Härdle (1990); Györfi et al. (2002); Tsybakov (2009).

Although our discussion below will assume a parametric function approximation method (and in many cases linear function approximation), many of the algorithms can be extended to nonparametric techniques. We will mention when such extensions exist as appropriate.

Up to now, the discussion implicitly assumed that the state is accessible for measurement. This is, however, rarely the case in practical applications. Luckily, the methods that we will discuss below do not actually need to access the states directly, but they can perform equally well when some "sufficiently descriptive feature-based representation" of the states is available (such as the camera images in the robot-arm example). A common way of arriving at such a representation is to construct

---

Algorithm 4 The function implementing the TD(λ) algorithm with linear function approximation. This function must be called after each transition.

function TDLAMBDALINFAPP(X, R, Y, θ, z)

Input: X is the last state, Y is the next state, R is the immediate reward associated with this transition, $\theta \in \mathbb{R}^d$ is the parameter vector of the linear function approximation, $z \in \mathbb{R}^d$ is the vector of eligibility traces

1: $\delta \leftarrow R + \gamma \cdot \theta^\top \varphi[Y] - \theta^\top \varphi[X]$

2: $z \leftarrow \varphi[X] + \gamma \cdot \lambda \cdot z$

3: $\theta \leftarrow \theta + \alpha \cdot \delta \cdot z$

4: **return** (θ, z)

---

state estimators (or observers, in control terminology) based on the history of the observations, which has a large literature both in machine learning and control. The discussion of these techniques, however, lies outside of the scope of the present paper.

2.2.1   TD(λ) WITH FUNCTION APPROXIMATION

Let us return to the problem of estimating a value function V of a Markov reward process $\mathcal{M} = (\mathcal{X}, \mathcal{P}_0)$, but now assume that the state space is large (or even infinite). Let $\mathcal{D} = ((X_t, R_{t+1}); t \geq 0)$ be a realization of $\mathcal{M}$. The goal, as before, is to estimate the value function of $\mathcal{M}$ given $\mathcal{D}$ in an incremental manner.

Choose a smooth parametric function-approximation method $(V_\theta; \theta \in \mathbb{R}^d)$ goal is to approximate the value function V underlying $\mathcal{M}$. (i.e., for any $\theta \in \mathbb{R}^d$, $V_\theta : \mathcal{X} \to \mathbb{R}$ is such that $\nabla_\theta V_\theta(x)$ exists for any $x \in \mathcal{X}$). The generalization of tabular TD(λ) with accumulating eligibility traces to the case when the value functions are approximated using members of $(V_\theta; \theta \in \mathbb{R}^d)$ uses the following updates (Sutton, 1984, 1988):

$$
\begin{aligned}
\delta_{t+1} &= R_{t+1} + \gamma V_{\theta_t}(X_{t+1}) - V_{\theta_t}(X_t), \\
z_{t+1} &= \nabla_\theta V_{\theta_t}(X_t) + \gamma \lambda z_t, \\
\theta_{t+1} &= \theta_t + \alpha_t \delta_{t+1} z_{t+1}, \\
z_0 &= 0.
\end{aligned}
\tag{2.5}
$$

Here $z_t \in \mathbb{R}^d$. Algorithm 4 shows the pseudocode of this algorithm.

To see that this algorithm is indeed a generalization of tabular TD(λ) assume that $\mathcal{X} = \{x_1, \ldots, x_D\}$ and let $V_\theta(x) = \theta^\top \varphi(x)$ with $\varphi_i(x) = \mathbb{I}_{\{x = x_i\}}$. Note that since V_θ is linear in the parameters (i.e., $V_\theta = \theta^\top \varphi$), it holds that $\nabla_\theta V_\theta = \varphi$. Hence, identifying $z_{t,i}$ ($\theta_{t,i}$) with $z_t(x_i)$ (resp., $\hat{V}_t(x_i)$) we see that the update (2.5), indeed, reduces to the previous one.

In the off-policy version of TD(λ), the definition of δ_{t+1} becomes

$$
\delta_{t+1} = R_{t+1} + \gamma V_{\theta_t}(Y_{t+1}) - V_{\theta_t}(X_t).
$$

Unlike the tabular case, under off-policy sampling, convergence is no longer be guaranteed, but, in fact, the parameters may diverge (see, e.g., Bertsekas and Tsitsiklis, 1996, Example 6.7, p. 307). This is true for linear function approximation when the distributions of $(X_t; t \geq 0)$ do not match the stationary distribution of the MRP $\mathcal{M}$. Another case when the algorithm may diverge is when it is used with a nonlinear function-approximation method (see, e.g., Bertsekas and Tsitsiklis, 1996, Example 6.6, p. 292). For further examples of instability, see Baird (1995); Boyan and Moore (1995).

On the positive side, almost sure convergence can be guaranteed when *(i)* a linear function-approximation method is used with $\varphi : \mathcal{X} \to \mathbb{R}^d$; *(ii)* the stochastic process $(X_t; t \geq 0)$ is an ergodic Markov process whose stationary distribution μ is the same as the stationary distribution of the MRP $\mathcal{M}$; and *(iii)* the step-size sequence satisfies the RM conditions (Tsitsiklis and Van Roy, 1997; Bertsekas and Tsitsiklis, 1996, p. 222, Section 5.3.7). In the results cited, it is also assumed that the components of φ (i.e., $\varphi_1, \ldots, \varphi_d$) are linearly independent. When this holds, the limit of the parameter vector will be unique. In the other case, i.e., when the features are redundant, the parameters will still converge, but the limit will depend on the parameter vector's initial value. However, the limiting value function will be unique (Bertsekas, 2010).

Assuming that TD(λ) converges, let $\theta^{(\lambda)}$. denote the limiting value of θ_t.

Let

$$\mathcal{F} = \{V_\theta \,|\, \theta \in \mathbb{R}^d\}$$

be the space of functions that can be represented using V_θ. Note that $\mathcal{F}$ is a linear subspace of the vector space of all real-valued functions with domain $\mathcal{X}$. The limit $\theta^{(\lambda)}$ is known to satisfy the so-called *projected fixed-point equation*

$$V_{\theta^{(\lambda)}} = \Pi_{\mathcal{F},\mu} T^{(\lambda)} V_{\theta^{(\lambda)}}, \tag{2.6}$$

where the operators $T^{(\lambda)}$ and $\Pi_{\mathcal{F},\mu}$ are defined as follows: For $m \in \mathbb{N}$ let $T^{[m]}$ be the *m-step lookahead Bellman operator*:

$$T^{[m]} \hat{V}(x) = \mathbb{E}\left[\sum_{t=0}^{m} \gamma^t R_{t+1} + \gamma^{m+1} \hat{V}(X_{m+1}) \,\Big|\, X_0 = x\right].$$

Clearly, V, the value function to be estimated is a fixed point of $T^{[m]}$ for any $m \geq 0$. Assume that $\lambda < 1$. Then, operator $T^{(\lambda)}$ is defined as the exponentially weighted average of $T^{[0]}, T^{[1]}, \ldots$:

$$T^{(\lambda)} \hat{V}(x) = (1 - \lambda) \sum_{m=0}^{\infty} \lambda^m T^{[m]} \hat{V}(x).$$

For $\lambda = 1$, we let $T^{(1)} \hat{V} = \lim_{\lambda \to 1-} T^{(\lambda)} \hat{V} = V$. Notice that for $\lambda = 0$, $T^{(0)} = T$. Operator $\Pi_{\mathcal{F},\mu}$ is a projection: It projects functions of states to the linear space $\mathcal{F}$ with respect to the weighted 2-norm $\|f\|_\mu^2 = \sum_{x \in \mathcal{X}} f^2(x)\mu(x)$:

$$\Pi_{\mathcal{F},\mu} \hat{V} = \underset{f \in \mathcal{F}}{\arg\min} \|\hat{V} - f\|_\mu.$$

The essence of the proof of convergence of TD(λ) is that the composite operator $\Pi_{\mathcal{F},\mu} T^{(\lambda)}$ is a contraction with respect to the norm $\|\cdot\|_\mu$. This result heavily exploits that μ is the stationary distribution underlying $\mathcal{M}$ (which defines $T^{(\lambda)}$). For other distributions, the composite operator might not be a contraction; in which case, TD(λ) might diverge.

As to the quality of the solution found, the following error bound holds for the fixed point of (2.6):

$$\left\| V_{\theta^{(\lambda)}} - V \right\|_\mu \le \frac{1}{\sqrt{1 - \gamma_\lambda}} \left\| \Pi_{\mathcal{F},\mu} V - V \right\|_\mu.$$

Here $\gamma_\lambda = \gamma(1 - \lambda)/(1 - \lambda\gamma)$ is the contraction modulus of $\Pi_{\mathcal{F},\mu} T^{(\lambda)}$ (Tsitsiklis and Van Roy, 1999a; Bertsekas, 2007b). (For sharper bounds, see Yu and Bertsekas 2008; Scherrer 2010.) From the error bound we see that $V_{\theta^{(1)}}$ is the best approximation to V within $\mathcal{F}$ with respect to the norm $\|\cdot\|_\mu$ (this should come at no surprise as TD(1) minimizes this mean-squared error by design). We also see that as we let $\lambda \to 0$ the bound allows for larger errors. It is known that this is not an artifact of the analysis. In fact, in Example 6.5 of the book by Bertsekas and Tsitsiklis (1996) (p. 288), a simple MRP with n states and a one-dimensional feature extractor φ is given such that $V_{\theta^{(0)}}$ is a very poor approximation to V, while $V_{\theta^{(1)}}$ is a reasonable approximation. Thus, in order to get good accuracy when working with $\lambda < 1$, it is *not* enough to choose the function space $\mathcal{F}$ so that the best approximation to V has small error. At this stage, however, one might wonder if using $\lambda < 1$ makes sense at all. A recent paper by Van Roy (2006) suggests that when considering performance loss bounds instead of approximation errors *and* the full control learning task (cf. Section 3), $\lambda = 0$ will in general be at *no disadvantage* compared to using $\lambda = 1$, at least, when state-aggregation is considered. Thus, while the mean-squared error of the solution might be large, when the solution is used in control, the performance of the resulting policy will still be as good as that of one that is obtained by calculating the TD(1) solution. However, the major reason to prefer TD(λ) with $\lambda < 1$ over TD(1) is because empirical evidence suggests that it converges much faster than TD(1), the latter of which, at least for practical sample sizes, often produces very poor estimates (e.g., Sutton and Barto, 1998, Section 8.6).

TD(λ) solves a model   Sutton et al. (2008) and Parr et al. (2008) observed independently of each other that the solution obtained by TD(0) can be thought of as the solution of a deterministic MRP with a linear dynamics. In fact, as we will argue now this also holds in the case of TD(λ). This suggests that if the deterministic MRP captures the essential features of the original MRP, $V_{\theta^{(\lambda)}}$ will be a good approximation to V. To firm up this statement, following Parr et al. (2008), let us study the *Bellman error*

$$\Delta^{(\lambda)}(\hat{V}) = T^{(\lambda)} \hat{V} - \hat{V}$$

of $\hat{V} : \mathcal{X} \to \mathbb{R}$ under $T^{(\lambda)}$. Note that $\Delta^{(\lambda)}(\hat{V}) : \mathcal{X} \to \mathbb{R}$. A simple contraction argument shows that $\left\| V - \hat{V} \right\|_\infty \le \frac{1}{1-\gamma} \left\| \Delta^{(\lambda)}(\hat{V}) \right\|_\infty$. Hence, if $\Delta^{(\lambda)}(\hat{V})$ is small, $\hat{V}$ is close to V.

The following error decomposition can be shown to hold:[2]

$$\Delta^{(\lambda)}(V_{\theta(\lambda)}) = (1 - \lambda) \sum_{m \geq 0} \lambda^m \Delta_m^{[r]} + \gamma \left\{ (1 - \lambda) \sum_{m \geq 0} \lambda^m \Delta_m^{[\varphi]} \right\} \theta^{(\lambda)}.$$

Here $\Delta_m^{[r]} = \bar{r}_m - \Pi_{\mathcal{F},\mu} \bar{r}_m$ and $\Delta_m^{[\varphi]} = P^{m+1} \varphi^\top - \Pi_{\mathcal{F},\mu} P^{m+1} \varphi^\top$ are the errors of modeling the m-step rewards and transitions with respect to the features φ, respectively; $\bar{r}_m : \mathcal{X} \to \mathbb{R}$ is defined by $\bar{r}_m(x) = \mathbb{E}\left[R_{m+1} \mid X_0 = x \right]$ and $P^{m+1} \varphi^\top$ denotes a function that maps states to d-dimensional row-vectors and which is defined by $P^{m+1} \varphi^\top(x) = (P^{m+1} \varphi_1(x), \ldots, P^{m+1} \varphi_d(x))$. Here $P^m \varphi_i : \mathcal{X} \to \mathbb{R}$ is the function defined by $P^m \varphi_i(x) = \mathbb{E}\left[\varphi_i(X_m) \mid X_0 = x \right]$. Thus, we see that the Bellman error will be small if the m-step immediate rewards and the m-step feature-expectations are well captured by the features. We can also see that as λ gets closer to 1, it becomes more important for the features to capture the structure of the value function, and as λ gets closer to 0, it becomes more important to capture the structure of the immediate rewards and the immediate feature-expectations. This suggests that the "best" value of λ (i.e., the one that minimizes $\|\Delta^{(\lambda)}(V_{\theta(\lambda)})\|$) may depend on whether the features are more successful at capturing the short-term or the long-term dynamics (and rewards).

2.2.2   GRADIENT TEMPORAL DIFFERENCE LEARNING

In Section 2.2.3, we will see some methods using which the issue of divergence of TD(λ) can be avoided. However, the computational (time and storage) complexity of these methods is significantly larger than that of TD(λ). In this section, we present two recent algorithms introduced by Sutton et al. (2009b,a), which also overcome the instability issue, converge to the TD(λ) solutions in the on-policy case, and yet they are almost as efficient as TD(λ).

For simplicity, we consider the case when $\lambda = 0$, $((X_t, R_{t+1}, Y_{t+1}); t \geq 0)$ is a stationary process, $X_t \sim \nu$ (ν can be different from the stationary distribution of $\mathcal{P}$) and when linear function approximation is used with linearly independent features. Assume that $\theta^{(0)}$, the solution to (2.6), exists. Consider the objective function

$$J(\theta) = \left\| V_\theta - \Pi_{\mathcal{F},\nu} T V_\theta \right\|_\nu^2. \tag{2.7}$$

Notice that all solutions to (2.6) are minimizers of J, and there are no other minimizers of J when (2.6) has solutions. Thus, minimizing J will give a solution to (2.6). Let θ_* denote a minimizer of J. Since, by assumption, the features are linearly independent, the minimizer of J is unique, i.e., θ_* is well-defined. Introduce the shorthand notations

$$\begin{aligned} \delta_{t+1}(\theta) &= R_{t+1} + \gamma V_\theta(Y_{t+1}) - V_\theta(X_t) \\ &= R_{t+1} + \gamma \theta^\top \varphi'_{t+1} - \theta^\top \varphi_t, \\ \varphi_t &= \varphi(X_t), \\ \varphi'_{t+1} &= \varphi(Y_{t+1}). \end{aligned} \tag{2.8}$$

---

[2]Parr et al. (2008) observed this for $\lambda = 0$. The extension to $\lambda > 0$ is new.

Algorithm 5 The function implementing the GTD2 algorithm. This function must be called after each transition.

---

function GTD2(X, R, Y, θ, w)

Input: X is the last state, Y is the next state, R is the immediate reward associated with this transition, $\theta \in \mathbb{R}^d$ is the parameter vector of the linear function approximation, $w \in \mathbb{R}^d$ is the auxiliary weight

1: $f \leftarrow \varphi[X]$
2: $f' \leftarrow \varphi[Y]$
3: $\delta \leftarrow R + \gamma \cdot \theta^\top f' - \theta^\top f$
4: $a \leftarrow f^\top w$
5: $\theta \leftarrow \theta + \alpha \cdot (f - \gamma \cdot f') \cdot a$
6: $w \leftarrow w + \beta \cdot (\delta - a) \cdot f$
7: **return** (θ, w)

A simple calculation allows us to rewrite J in the following form:

$$J(\theta) = \mathbb{E}\left[\delta_{t+1}(\theta)\varphi_t\right]^\top \mathbb{E}\left[\varphi_t \varphi_t^\top\right]^{-1} \mathbb{E}\left[\delta_{t+1}(\theta)\varphi_t\right]. \tag{2.9}$$

Taking the gradient of the objective function we get

$$\nabla_\theta J(\theta) = -2\mathbb{E}\left[(\varphi_t - \gamma\varphi'_{t+1})\varphi_t^\top\right] w(\theta), \tag{2.10}$$

where

$$w(\theta) = \mathbb{E}\left[\varphi_t\varphi_t^\top\right]^{-1} \mathbb{E}\left[\delta_{t+1}(\theta)\varphi_t\right].$$

Let us now introduce two sets of weights: θ_t to approximate θ_* and w_t to approximate $w(\theta_*)$. In GTD2 ("gradient temporal difference learning, version 2"), the update of θ_t is chosen to follow the negative stochastic gradient of J based on (2.10) assuming that $w_t \approx w(\theta_t)$, while the update of w_t is chosen so that for any fixed θ, w_t would converge almost surely to $w(\theta)$:

$$\theta_{t+1} = \theta_t + \alpha_t (\varphi_t - \gamma\varphi'_{t+1}) \varphi_t^\top w_t,$$
$$w_{t+1} = w_t + \beta_t (\delta_{t+1}(\theta_t) - \varphi_t^\top w_t) \varphi_t.$$

Here $(\alpha_t; t \geq 0)$, $(\beta_t; t \geq 0)$ are two step-size sequences. Note that the update equation for $(w_t; t \geq 0)$ is just the basic Least-Mean Square (LMS) rule, which is a widely used update rule in signal processing (Widrow and Stearns, 1985). Sutton et al. (2009a) have shown that under the standard RM conditions on the step-sizes and some other mild technical conditions (θ_t) converges to the minimizer of $J(\theta)$, almost surely. However, unlike for TD(0), convergence is guaranteed independently of the distribution of $(X_t; t \geq 0)$. At the same time, the update of GTD2 costs only twice as much as the cost of TD(0). Algorithm 5 shows the pseudocode of GTD2.

To arrive at the second algorithm called TDC ("temporal difference learning with corrections"), write the gradient as

$$\nabla_\theta J(\theta) = -2\Big(\mathbb{E}\big[\delta_{t+1}(\theta)\varphi_t\big] - \gamma\mathbb{E}\big[\varphi'_{t+1}\varphi_t^\top\big]w(\theta)\Big).$$

Leaving the update w_t unchanged, we then arrive at,

$$\begin{aligned}
\theta_{t+1} &= \theta_t + \alpha_t\left(\delta_{t+1}(\theta_t)\varphi_t - \gamma\varphi'_{t+1}\,\varphi_t^\top w_t\right),\\
w_{t+1} &= w_t + \beta_t\left(\delta_{t+1}(\theta_t) - \varphi_t^\top w_t\right)\varphi_t.
\end{aligned}$$

The pseudocode of this update is identical to that of GTD2 except that line 5 should be replaced by

$$\theta \leftarrow \theta + \alpha\cdot(\delta\cdot f - \gamma\cdot a\cdot f').$$

In TDC, the update of w_t must use larger step-sizes than the update of θ_t: $\alpha_t = o(\beta_t)$. This makes TDC a member of the family of the so-called *two-timescale stochastic approximation algorithms* (Borkar, 1997, 2008). If, in addition to this condition, the standard RM conditions are also satisfied by both step-size sequences, $\theta_t \rightarrow \theta_*$ holds again almost surely (Sutton et al., 2009a). More recently, these algorithms have been extended to nonlinear function approximation (Maei et al., 2010a). Also, one can show that it suffices if $\alpha_t \ll \beta_t$ (Maei, 2010, personal communication). The algorithms can also be extended to use eligibility traces (Maei and Sutton, 2010).

Note that although these algorithms are derived from the gradient of an objective function, they are not true stochastic gradient methods in the sense that the expected weight update direction can be different from the direction of the negative gradient of the objective function. In fact, these methods belong to the larger class of pseudo-gradient methods. The two methods differ in how they approximate the gradients, and it remains to be seen whether one of them is better than the other.

2.2.3   LEAST-SQUARES METHODS

The methods discussed so far are similar to the LMS algorithm of adaptive filtering in that they are taking small steps in the parameter space following some noisy, gradient-like signal. As such, similarly to the LMS algorithm, they are sensitive to the choice of the step-sizes, the distance between the initial parameter and the limit point $\theta^{(\lambda)}$, or the eigenvalue structure of the matrix A that determines the dynamics of updates (e.g., for TD(0), $A = \mathbb{E}\big[\varphi_t(\varphi_t - \gamma\varphi'_{t+1})^\top\big]$). Over the years, many ideas appeared in the literature to address these issues. These are essentially parallel to those available in the adaptive filtering literature. A non-exhaustive list includes the use of adaptive step-sizes (Sutton, 1992; George and Powell, 2006), normalizing the updates (Bradtke, 1994) or reusing previous samples (Lin, 1992). Although these techniques can indeed help, each have their own weaknesses. In adaptive filtering, the algorithm that is known to address all the deficiencies of LMS is known as the LS ("least-squares") algorithm. In this section, we review the analogous methods of reinforcement learning.

LSTD: Least-squares temporal difference learning   In the limit of an infinite number of examples, TD(0) finds a parameter vector θ that satisfies

$$\mathbb{E}\left[\varphi_t\,\delta_{t+1}(\theta)\right] = 0, \tag{2.11}$$

where we used the notation of the previous section. Given a finite sample

$$\mathcal{D}_n = ((X_0, R_1, Y_1), (X_1, R_2, Y_2), \ldots, (X_{n-1}, R_n, Y_n)),$$

one can approximate (2.11) by

$$\frac{1}{n}\sum_{t=0}^{n-1}\varphi_t\,\delta_{t+1}(\theta) = 0. \tag{2.12}$$

Plugging in $\delta_{t+1}(\theta) = R_{t+1} - (\varphi_t - \gamma\varphi'_{t+1})^{\top}\theta$, we see that this equation is linear in θ. In particular, if the matrix $\hat{A}_n = \frac{1}{n}\sum_{t=0}^{n-1}\varphi_t(\varphi_t - \gamma\varphi'_{t+1})^{\top}$ is non-singular, the solution is simply

$$\theta_n = \hat{A}_n^{-1}\hat{b}_n, \tag{2.13}$$

where $\hat{b}_n = \frac{1}{n}\sum_{t=0}^{n-1}R_{t+1}\varphi_t$. If inverting $\hat{A}_n$ can be afforded (i.e., the dimensionality of the features is not too large and the method is not called too many times) then this method can give a better approximation to the equilibrium solution than TD(0) or some other incremental first-order method since the latter are negatively impacted by the eigenvalue spread of the matrix $\mathbb{E}\left[\hat{A}_n\right]$.

The idea of directly computing the solution of (2.12) is due to Bradtke and Barto (1996), who call the resulting algorithm *least-squares temporal difference* learning or LSTD. Using the terminology of stochastic programming, LSTD can be seen to use *sample average approximation* (Shapiro, 2003). In the terminology of statistics, it belongs to the so-called Z-estimation family of procedures (e.g., Kosorok, 2008, Section 2.2.5). It is a simple observation that when the LSTD solution exists, LSTD minimizes the empirical approximation to the projected squared Bellman error, $\left\|\Pi_{\mathcal{F},\mu}(TV - V)\right\|_{\mu}^{2}$, over the linear space $\mathcal{F}$ (Antos et al., 2008).

Using the Sherman-Morrison formula, one can derive an incremental version of LSTD, analogously to how the recursive least-squares (RLS) method is derived in adaptive filtering Widrow and Stearns (1985). The resulting algorithm is called "recursive LSTD" (RLSTD) and works as follows (Bradtke and Barto, 1996): Choose $\theta_0 \in \mathbb{R}^d$ and let $C_0 \in \mathbb{R}^{d\times d}$ such that C_0 is a "small" positive definite matrix (e.g., $C_0 = \beta I$, for $\beta > 0$ "small"). Then, for $t \geq 0$,

$$C_{t+1} = C_t - \frac{C_t\,\varphi_t(\varphi_t - \gamma\varphi'_{t+1})^{\top}C_t}{1 + (\varphi_t - \gamma\varphi'_{t+1})^{\top}C_t\varphi_t},$$

$$\theta_{t+1} = \theta_t + \frac{C_t}{1 + (\varphi_t - \gamma\varphi'_{t+1})^{\top}C_t\varphi_t}\,\delta_{t+1}(\theta_t)\varphi_t.$$

The computational complexity of one update is $O(d^2)$. Algorithm 6 shows the pseudocode of this algorithm.

Algorithm 6 The function implementing the RLSTD algorithm. This function must be called after each transition. Initially, C should be set to a diagonal matrix with small positive diagonal elements: $C = \beta\, I$, with $\beta > 0$.

---

function RLSTD(X, R, Y, C, θ)

Input: X is the last state, Y is the next state, R is the immediate reward associated with this transition, $C \in \mathbb{R}^{d \times d}$, and $\theta \in \mathbb{R}^d$ is the parameter vector of the linear function approximation

1: $f \leftarrow \varphi[X]$
2: $f' \leftarrow \varphi[Y]$
3: $g \leftarrow (f - \gamma f')^{\top} C$ $\triangleright$ g is a $1 \times d$ row vector
4: $a \leftarrow 1 + gf$
5: $v \leftarrow Cf$
6: $\delta \leftarrow R + \gamma \cdot \theta^{\top} f' - \theta^{\top} f$
7: $\theta \leftarrow \theta + \delta / a \cdot v$
8: $C \leftarrow C - v g / a$
9: **return** (C, θ)

Boyan (2002) extended LSTD to incorporate the λ parameter of TD(λ) and called the resulting algorithm LSTD(λ). (Note that for $\lambda > 0$ to make sense one needs $X_{t+1} = Y_{t+1}$; otherwise, the TD errors do not telescope). The LSTD(λ) solution is derived from (2.5). It is defined as the parameter value that makes the cumulated updates zero:

$$\frac{1}{n} \sum_{t=0}^{n-1} \delta_{t+1}(\theta) z_{t+1} = 0, \tag{2.14}$$

where $z_{t+1} = \sum_{s=0}^{t} (\gamma \lambda)^{t-s} \varphi_s$ are the eligibility traces. This is again linear in θ and the previous comments apply. The recursive form of LSTD(λ), RLSTD(λ), has been studied by Xu et al. (2002) and (independently) by Nedič and Bertsekas (2003). (See Algorithm 16 for the pseudocode of a closely related algorithm.)

One issue with LSTD(λ) as stated here that Equation (2.14) might fail to have a solution. In the on-policy case, for large enough sample sizes at least, a solution will always exist. When a solution does not exist, a commonly suggested trick is to add a small positive diagonal matrix to the matrix to be inverted (this corresponds to starting with a diagonal matrix in RLSTD). However, this trick is not guaranteed to work. A better approach is based on the observation that when the matrix is invertible then the LSTD parameter vector is a minimizer of the projected Bellman error. Since the minimizer of the projected Bellman error is always well defined, instead of the solving for the zero of (2.14) , one can aim for minimizing the projected Bellman error.

Under standard assumptions on the sample, it follows from the law of large numbers and a simple continuity argument that LSTD(λ) (and its recursive variants) converge almost surely to the solution of the projected fixed-point equation (2.6) if this solution exists. This was formally shown for $\lambda = 0$ by Bradtke and Barto (1996), and for $\lambda > 0$ by Xu et al. (2002) and Nedič and Bertsekas (2003). Although these results were shown only for the on-policy case, it is easy to see that they also hold in the off-policy case *provided* that the limiting solution exists.

As promised, (R)LSTD(λ) avoids the difficulties associated with tuning the incremental algorithms: It neither relies on step-sizes, nor is it sensitive to the eigenvalue structure of A, or the choice of the initial value of θ. Experimental results by Bradtke and Barto (1996); Boyan (2002); Xu et al. (2002) and others have indeed confirmed that the parameters obtained using (R)LSTD(λ) converge faster than those obtained by TD(λ). However, their computational properties are quite different from those of TD(λ). We will discuss the implications of this after we have reviewed the LSPE algorithm.

LSPE: Least-squares policy evaluation   An alternative to LSTD (and LSTD(λ)) is λ-*least squares policy evaluation* (λ-LSPE for short) due to Bertsekas and Ioffe (1996). The basic idea of this algorithm is to mimic *multi-step value iteration*. Again, the method assumes that linear function-approximation is used.

It works as follows. Define the $(n - s)$-step prediction of the value of X_s as

$$\hat{V}_{s,n}^{(\lambda)}(\theta) = \theta^\top \varphi_s + \sum_{q=s}^{n-1} (\gamma\lambda)^{q-s} \delta_{q+1}(\theta)$$

and define the loss

$$J_n(\hat{\theta}, \theta) = \frac{1}{n} \sum_{s=0}^{n-1} \left(\hat{\theta}^\top \varphi_s - \hat{V}_{s,n}^{(\lambda)}(\theta) \right)^2 .$$

Then, λ-LSPE updates the parameters by

$$\theta_{t+1} = \theta_t + \alpha_t (\underset{\hat{\theta}}{\operatorname{argmin}} \, J_{n_t}(\hat{\theta}, \theta_t) - \theta_t), \qquad (2.15)$$

where $(\alpha_t; t \geq 0)$ is a step-size sequence and $(n_t; t \geq 0)$ is a non-decreasing sequence of integers. (Bertsekas and Ioffe (1996) only considered the case when $n_t = t$, which is a logical choice when the algorithm is used in an online learning scenario. When the algorithm is used with a finite (say, n) observations, we can set $n_t = n$ or $n_t = \min(n, t)$.) Note that J_n is quadratic in $\hat{\theta}$, hence the solution to the minimization problem can be obtained in closed form. The resulting algorithm is shown as Algorithm 7. A recursive, incremental version of λ-LSPE is also available. Similarly to LSTD(λ), it requires $O(d^2)$ operations per time step when $n_t = t$.

To get a sense of the behavior of λ-LSPE, consider the update in the special case when $\lambda = 0$ and $\alpha_t = 1$:

$$\theta_{t+1} = \underset{\hat{\theta}}{\operatorname{argmin}} \, \frac{1}{n_t} \sum_{s=0}^{n_t-1} \left\{ \hat{\theta}^\top \varphi(X_s) - (R_{s+1} + \gamma V_{\theta_t}(Y_{s+1})) \right\}^2 .$$

Algorithm 7 The function implementing the batch-mode λ-LSPE update. This function must be called repeatedly until convergence.

---

function LAMBDALSPE(D, θ)
Input: $D = ((X_t, A_t, R_{t+1}, Y_{t+1}); t = 0, \ldots, n-1)$ is a list of transitions, $\theta \in \mathbb{R}^d$ is the parameter vector
1: $A, b, \delta \leftarrow 0$ $\triangleright A \in \mathbb{R}^{d \times d}, b \in \mathbb{R}^d, \delta \in \mathbb{R}$
2: **for** $t = n-1$ **downto** 0 **do**
3:      $f \leftarrow \varphi[X_t]$
4:      $v \leftarrow \theta^\top f$
5:      $\delta \leftarrow \gamma \cdot \lambda \cdot \delta + \left(R_{t+1} + \gamma \cdot \theta^\top \varphi[Y_{t+1}] - v\right)$
6:      $b \leftarrow b + (v + \delta) \cdot f$
7:      $A \leftarrow A + f \cdot f^\top$
8: **end for**
9: $\theta' \leftarrow A^{-1} b$
10: $\theta \leftarrow \theta + \alpha \cdot (\theta' - \theta)$
11: **return** θ

Thus, in this case, λ-LSPE solves a linear regression problem, implementing the so-called *fitted value iteration* algorithm for policy evaluation with linear function approximation. For a fixed, non-random value of θ_t, the true regression function underlying the above least-squares problem is $\mathbb{E}\left[R_{s+1} + \gamma V_{\theta_t}(Y_{s+1}) | X_s = x\right]$, which is just $T V_{\theta_t}(x)$. Thus, if the function space $\mathcal{F}$ is rich enough and the sample size n_t is large, one may expect $\theta_{t+1}^\top \varphi$ to be close to $T V_{\theta_t}(x)$, and we see that the algorithm implements value iteration in an approximate manner. The case when $\lambda > 0$ can be given a similar interpretation.

When $\alpha_t < 1$, the parameters are moved towards the minimizer of $J_{n_t}(\cdot, \theta_t)$ in proportion to the size of α_t. The role of smoothing the updates this way is *(i)* to stabilize the parameters for small sample sizes (i.e., when n_t and d are in the same range) and *(ii)* to ensure that policies are changed gradually when the algorithm is used as a subroutine of a control algorithm (cf. Section 3). The idea of smoothing the parameter updates could also be used together with LSTD.

Just like LSTD(λ), the multi-step version of λ-LSPE (i.e., when $\lambda > 0$) requires $X_{t+1} = Y_{t+1}$. The parameter λ plays a role similar to its role in other TD methods: Increasing λ is expected to reduce bias and increase variance, though unlike TD(λ), λ-LSPE bootstraps even when $\lambda = 1$. However, the effect of bootstrapping is diminishing with $n_t \to \infty$.

Under standard assumptions on the sample and when $n_t = t$, λ-LSPE is known to converge almost surely to the solution of the projected fixed-point equation (2.6), both for decreasing (Nedić and Bertsekas, 2003) and constant step-sizes (Bertsekas et al., 2004). In the latter case, convergence is guaranteed if $0 < \alpha_t \equiv \alpha < (2 - 2\gamma\lambda)/(1 + \gamma - 2\gamma\lambda)$. Note that 1 is always included in this range.

According to Bertsekas et al. (2004) λ-LSPE is competitive with LSTD in the sense that the distance between the parameters updated by LSTD(λ) and λ-LSPE becomes, very soon, smaller than the statistical inaccuracy, resulting from the use of a finite sample. Experimental results obtained by Bertsekas et al. (2004) and earlier by Bertsekas and Ioffe (1996) to train a Tetris playing program indicate that λ-LSPE is, indeed, a competitive algorithm. Moreover, λ-LSPE is always well-defined (all inverses involved exist in the limit or with appropriate initialization), whereas LSTD(λ) might be ill-defined in off-policy settings.

Comparing least-squares and TD-like methods. The price of the increased stability and accuracy of least-squares techniques is their increased computational complexity. In particular, for a sample of size n, the complexity of a straightforward implementation of LSTD is $O(nd^2 + d^3)$, while the complexity of RLSTD is $O(nd^2)$ (the same applies to LSPE). For comparison, the computational complexity of the lightweight, incremental methods discussed previously is only $O(nd)$ (or less when the features are sparse). Thus, the lightweight algorithms can do d passes on the sample while a least-squares method makes a single pass. The trick of saving and reusing the observations to increase the accuracy of TD-based algorithms was first suggested by Lin (1992), who dubbed his method "experience replay". When the value of d is large, this may be enough for the lightweight methods to perform as well as the least-squares methods given the same computation time. When d is very large, the least-squares methods might not be feasible at all. For example, Silver et al. (2007) use over a million features when building a value function for the game of Go. When d is in this range, least-squares methods are not feasible.

It becomes very complicate to compare these approaches if we take into account the frequency at which the observations arrive, the storage space available, the access time of storage, etc. Hence, here we look at one interesting case when new observations are available at negligible cost. In this case, there is no need to (store and) reuse data and the quality of solutions will depend on the methods' computation speed.

To compare the two approaches, fix some time T available for computation. In time T, the least-squares methods are limited to process a sample of size $n \approx T/d^2$, while the lightweight methods can process a sample of size $n' \approx nd$. Let us now look at the precision of the resulting parameters. Assume that the limit of the parameters is θ_*. Denote by θ_t the parameter obtained by (say) LSTD after processing t observations and denote by θ_t' the parameter obtained by a TD-method. Then, one expects that $\|\theta_t - \theta_*\| \approx C_1 t^{-\frac{1}{2}}$ and $\|\theta_t' - \theta_*\| \approx C_2 t^{-\frac{1}{2}}$. Thus,

$$\frac{\|\theta_{n'}' - \theta_*\|}{\|\theta_n - \theta_*\|} \approx \frac{C_2}{C_1} d^{-\frac{1}{2}}. \tag{2.16}$$

Hence, if $C_2/C_1 < d^{1/2}$ then the lightweight TD-like method will achieve a better accuracy, while in the opposite case the least-squares procedures will perform better. As usual, it is difficult to decide this *a priori*. As a rule of thumb, based on (2.16), we expect that when d is relatively small, least-squares methods might be converging faster; while if d is large, then the lightweight, incremental methods will give better results given a fixed computation budget. Notice that this analysis is not specific to reinforcement learning methods, but it applies in all cases when an incremental lightweight procedure is compared to a least-squares-like procedure (for a similar analysis in a supervised learning problem see, e.g., Bottou and Bousquet, 2008).

Realizing the need for efficient and robust methods, Geramifard et al. (2007) have recently introduced an incremental version of LSTD, called iLSTD, which, just like LSTD, computes the matrix $\hat{A}_n$ and vector $\hat{b}_n$, but in each time step, only one dimension of the parameter vector is updated. For sparse features, i.e., when only s components of the feature vector are nonzero, the per-iteration complexity of this method is $O(sd)$, while experimentally it has been demonstrated that it is almost as accurate as LSTD given the same number of samples (assuming again sparsity). The storage space needed by iLSTD after processing n samples is $O(\min(ns^2 + d, d^2))$. Thus, when the features are sparse and $ns^2 \ll d^2$, iLSTD might be competitive with LSTD and incremental TD-methods.

2.2.4   THE CHOICE OF THE FUNCTION SPACE

In order to be able to discuss the choice of the function space in a meaningful manner, we need to define how the quality of an approximate value function is measured. When the ultimate goal is value-prediction, a reasonable choice is to use the mean-squared error (MSE) with respect to an appropriate distribution over the states (say, μ). The choice of a metric is less clear when the goal is to learn a good controller and value estimation is only used as a subroutine of a more complex algorithm (such as the ones that will be reviewed in Section 3). Therefore, in lack of a good understanding of this case, for the purpose of this section, we will stick to the MSE as the quality-measure. However, we believe that most of the conclusions of this section would hold for other measures, too.

Learning can be viewed as the process of selecting some function from some space of functions ($\mathcal{F}$) that can be represented (finitely) in the computer's memory.[3] For simplicity, assume that the functions available are described by d parameters: $\mathcal{F} = \{V_\theta \mid \theta \in \mathbb{R}^d\}$. One measure that characterizes the choice of $\mathcal{F}$ is how well functions from $\mathcal{F}$ can approximate the target function V, leading to the definition of the *approximation error* underlying $\mathcal{F}$:

$$\inf_{V_\theta \in \mathcal{F}} \|V_\theta - V\|_\mu.$$

To decrease the approximation error, one is encouraged to choose a larger function space (i.e., when V_θ is linear we may add independent features to make $\mathcal{F}$ larger). However, as we will argue now, since learning by definition uses incomplete information, increasing the size of the function space is a double-edged sword.

For simplicity, let us consider linear function approximation, and assume that LSTD, as specified by (2.12), is used to obtain the parameters. To make the situation even simpler, assume that the discount factor, γ, is zero and $((X_t, R_{t+1}); t \geq 0)$ is an i.i.d. sample with $X_t \sim \mu$. In this case,

$$V(x) = r(x) = \mathbb{E}\left[R_{t+1}|X_t = x\right].$$

---

[3]We do not deal with issues of precision, i.e., that computers cannot really represent real numbers.

Thanks to $\gamma = 0$, LSTD can actually be seen to compute the minimizer of the *empirical loss* function,

$$L_n(\theta) = \frac{1}{n}\sum_{t=0}^{n-1}(\theta^\top \varphi(X_t) - R_{t+1})^2.$$

Assume that the dimensionality of the feature space, d, is so large that the matrix whose rows are $\varphi(X_0)^\top, \ldots, \varphi(X_{n-1})^\top$ has full column rank (in particular, assume that $d \geq n$). This implies that the minimum of L_n is zero and if θ_n denotes the solution of (2.12) then $\theta_n^\top \varphi(X_t) = R_{t+1}$ holds for $t = 0, \ldots, n-1$. If the observed rewards are noisy, the resulting function will be a poor approximation to the value function, V, i.e., the *estimation error*, $\|\theta_n^\top \varphi - V\|_\mu$, will be large. The phenomenon of fitting to the "noise" is called *overfitting*. If a smaller d is chosen (in general, if a smaller function space $\mathcal{F}$ is chosen), then overfitting will be less likely to happen. However, in this case, the approximation error will get larger. Hence, there is a tradeoff between the approximation and the estimation errors.

To quantify this tradeoff, let θ_* be the parameter vector that minimizes the loss

$$L(\theta) = \mathbb{E}\left[(\theta^\top \varphi(X_t) - R_{t+1})^2\right].$$

That is,

$$\theta_* = \operatorname*{argmin}_\theta L(\theta).$$

(A simple argument shows that V_{θ_*} is actually the projection of V to $\mathcal{F}$.) The following bound is known to hold if the random rewards are bounded by some value $\mathcal{R}$ *and*, after finding the optimal weights, at prediction time, the predicted values are back-projected to $[-\mathcal{R}, \mathcal{R}]$ (Györfi et al., 2002, Theorem 11.3, p. 192):

$$\mathbb{E}\left[\|\theta_n^\top \varphi - V\|^2\right] \leq C_1 \frac{d(1 + \log n)}{n} + C_2 \|\theta_*^\top \varphi - r\|^2. \tag{2.17}$$

Here C_2 is a universal constant, while C_1 is a constant that scales linearly with the variance and range of the random rewards.[4] The first term on the right-hand side bounds the estimation error, while the second term is due to the approximation error. Increasing d increases the first term, while it is generally expected to decrease the second.

The argument that leads to bounds of the above form is as follows: By the law of large numbers, $L_n(\theta)$ converges to $L(\theta)$ for any *fixed* value of θ. Hence, by minimizing $L_n(\theta)$, one hopes to obtain a good approximation to θ_* (more precisely, to $\theta_*^\top \varphi$). However, that $L_n(\theta)$ converges to $L(\theta)$ at every value of θ does not mean that the function $L_n(\cdot)$ is *uniformly close* to $L(\cdot)$. Thus, the minimizer of L_n might not give a small loss, as measured by L (cf. Figure 2.2). Guaranteeing that the two functions are uniformly close to each other (say, over the set $\{\theta \mid L_n(\theta) \leq L_n(0)\}$) is harder when the dimensionality of θ is larger, hence the tradeoff between estimation and approximation errors.

---

[4]Note that without truncation, C_1 would not be independent of the distribution of X_t. Although the theorem is stated for the expected error, similar results can be shown to hold with high probability.

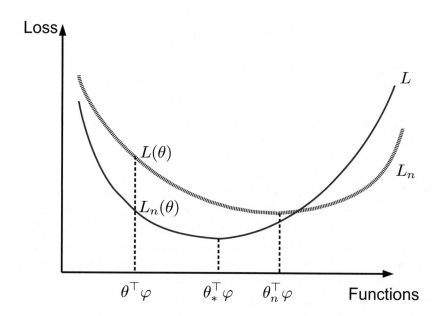

Figure 2.2: Convergence of $L_n(\cdot)$ to $L(\cdot)$. Shown are the curves of the empirical loss, L_n and the true loss L, as a function of the parameter θ. If the true curves are uniformly close to each other (i.e., for every θ, $L_n(\theta) - L(\theta)$ is small), then one can expect that the loss of $\theta_n^\top \varphi$ will be close to the loss of $\theta_*^\top \varphi$.

Bounds similar to (2.17) hold even when $\gamma > 0$, e.g. for value functions estimated using LSTD and even when the sequence $(X_t; t \geq 0)$ is dependent provided that it "mixes well" (for some initial steps in this direction consult the work of Antos et al. (2008)). In fact, when $\gamma \neq 0$, the noise comes both from the immediate rewards R_{t+1} and the "next states", Y_{t+1}. The trade-off between the approximation and estimation errors also shows up when control algorithms are used: Munos and Szepesvári (2008) and Antos et al. (2007, 2008) derive finite-sample performance bounds for some variants of fitted value iteration, a fitted actor-critic method and an approximate policy iteration method, respectively.

Recognizing the importance of the choice of the function space, as well as the difficulty of choosing it right, there has been a growing interest in automating this choice lately. One class of methods aims at constructing a parsimonious set of features (basis functions). These include tuning the parameter of Gaussian RBF either using a gradient- or the cross-entropy-method in the context of LSTD (Menache et al., 2005), deriving new basis functions with nonparametric techniques

(Keller et al., 2006; Parr et al., 2007) or using a combination of numerical analysis and nonparametric techniques (Mahadevan, 2009). These methods, however, do not attempt to control the tradeoff between the approximation and estimation errors. To account for this deficiency, other researchers explore nonparametric techniques originating at supervised learning. Examples of this line of research include the use of regression trees (Ernst et al., 2005), or "kernelizing" the value estimation algorithms (e.g., Rasmussen and Kuss, 2004; Engel et al., 2005; Ghavamzadeh and Engel, 2007; Xu et al., 2007; Taylor and Parr, 2009). These approaches implicitly or explicitly regularize the estimates to control the loss. Kolter and Ng (2009) designed an algorithm inspired by LASSO that uses ℓ^1-regularization to implement feature selection in the context of LSTD. Although the approaches above are inspired by principled methods of supervised learning, not much is known about their statistical properties. Recently, Farahmand et al. (2009, 2008) have developed another regularization-based approach that comes with statistical guarantees.

The difficulty of using (some) nonparametric techniques is that they are computationally expensive. As a result, when the algorithms are used for planning and a fast simulator is used to generate data (so that the cost of generating new data is negligible), it might be better to use an appropriate fast incremental method also act as a way of regularization) and a simple linear function-approximation method with many features than to use a sophisticated but computationally expensive nonparametric method. Computational efficiency is less important if a limited amount of data is available only and the quality of the solutions is the primary concern, or when the problem is complex enough so that tuning the function approximation is necessary, but hand-tuning is infeasible.

CHAPTER 3

Control

We now turn to the problem of learning a (near-)optimal policy. We start by discussing the various forms of control learning problems (Section 3.1), followed by a discussion of interactive learning (Section 3.2). In the last two sections (Sections 3.3 and 3.4), the learning counterparts of the classical methods of dynamic programming are discussed.

3.1   A CATALOG OF LEARNING PROBLEMS

Figure 3.1 shows the basic types of control learning problems. The first criterion that the space of problems is split upon is whether the learner can actively influence the observations. In case she can, then we talk about *interactive learning*, otherwise one is facing a *non-interactive learning* problem.[1] Interactive learning is potentially easier since the learner has the additional option to influence the distribution of the sample. However, the goal of learning is usually different in the two cases, making these problems incomparable in general.

In the case of non-interactive learning, the natural goal is to find a good policy given the observations. A common situation is when the sample is fixed. For example, the sample can be the result of some experimentation with some physical system that happened before learning started. In machine learning terms, this corresponds to *batch learning*. (Batch learning problems are not to be confused with batch learning methods, which are the opposite of incremental a.k.a. recursive, or iterative methods.) Since the observations are uncontrolled, the learner working with a fixed sample

[1]The terms "active learning" and "passive learning" might appeal and their meaning indeed covers the situations discussed here. However, unfortunately, the term "active learning" is already reserved in machine learning for a special case of interactive learning. As a result, we also decided against calling non-interactive learning "passive learning" so that no one is tempted to call interactive learning "active learning".

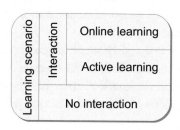

Figure 3.1: Types of reinforcement problems

has to deal with an off-policy learning situation. In other cases, the learner can ask for more data (i.e., when a simulator is used to generate new data). Here the goal might be to learn a good policy as quickly as possible.

Consider now interactive learning. One possibility is that learning happens while interacting with a real system in a closed-loop fashion. A reasonable goal then is to optimize *online performance*, making the learning problem an instance of *online learning*. Online performance can be measured in different ways. A natural measure is to use the sum of rewards incurred during learning. An alternative cost measure is the number of times the learner's future expected return falls short of the optimal return, i.e., the number of times the learner commits a "mistake". Another possible goal is to produce a well-performing policy as soon as possible (or find a good policy given a finite number of samples), just like in non-interactive learning. As opposed to the non-interactive situation, however, here the learner has the option to control the samples so as to maximize the chance of finding such a good policy. This learning problem is an instance of *active learning*.

When a simulator is available, the learning algorithms can be used to solve *planning problems*. In planning the previous performance metrics become irrelevant and the algorithms' running time and memory requirements become the primary concern.

3.2   CLOSED-LOOP INTERACTIVE LEARNING

The special feature of interactive learning is the need to *explore*. In this section, we first use bandits (i.e., MDPs with a single state) to illustrate the need for exploration, both in online and active learning. Next, we discuss active learning in MDPs. This is followed by a discussion of algorithms available for online learning in MDPs.

3.2.1   ONLINE LEARNING IN BANDITS

Consider an MDP that has a single state. Let the problem be that of maximizing the return while learning. Since there is only one state, this is an instance of the classical *bandit problems* (Robbins, 1952). A basic observation then is that a bandit learner who always chooses the action with the best estimated payoff (i.e., who always makes the *greedy* choice) can fail to find the best action with positive probability, which in turn leads to a large loss. Thus, a good learner must take actions that look suboptimal, i.e., must *explore*. The question is then how to balance the frequency of exploring and exploiting (i.e., greedy) actions.

A simple strategy is to fix $\varepsilon > 0$ and choose a randomly selected action with probability ε, and go with the greedy choice otherwise. This is the so-called ε-*greedy* strategy. Another simple strategy is the so-called "Boltzmann exploration" strategy, according to which, given the sample means, $(Q_t(a); a \in \mathcal{A})$, of the actions at time t, the next action is drawn from the multinomial distribution $(\pi(a); a \in \mathcal{A})$, where

$$\pi(a) = \frac{\exp(\beta \, Q_t(a))}{\sum_{a' \in \mathcal{A}} \exp(\beta \, Q_t(a'))} \, .$$

Here $\beta > 0$ controls the greediness of action selection ($\beta \to \infty$ results in a greedy choice). The difference between Boltzmann exploration and ε-greedy is that ε-greedy does not take into account the relative values of the actions, while Boltzmann exploration does. These algorithms extend easily to the case of unrestricted MDPs provided that some estimates of the action-values is available.

If the parameter of ε-greedy is made a function of time and the resulting sequence is appropriately tuned, ε-greedy can be made competitive with other, more sophisticated algorithms. However, the best choice is problem dependent and there is no known automated way of obtaining good results with ε-greedy (Auer et al., 2002). The same holds for the Boltzmann exploration strategy.

A better approach might be to implement the so-called *optimism in the face of uncertainty* (OFU) principle due to Lai and Robbins (1985), according to which the learner should choose the action with the best upper confidence bound (UCB). A very successful recent algorithm, UCB1, implements this principle by assigning the following UCB to action a at time t (Auer et al., 2002):

$$U_t(a) = r_t(a) + \mathcal{R}\sqrt{\frac{2\log t}{n_t(a)}}.$$

Here $n_t(a)$ is the number of times action a was selected up to time t and $r_t(a)$ is the sample mean of the $n_t(a)$ rewards observed for action a, whose range is $[-\mathcal{R}, +\mathcal{R}]$. It can be shown that the failure probability of $U_t(a)$ is t^{-4}. Notice that an action's UCB is larger if less information is available for it. Further, an action's UCB value increases even if it is not tried. Algorithms 8 and 9 show the pseudocode of UCB1, in the form of two routines, one to be used for action selection and the other for updating the internal statistics.

When the variance of the rewards associated with some of the actions are small, it makes sense to estimate these variances and use them in place of the range $\mathcal{R}$ in the above algorithm. A principled way of doing this was proposed and analyzed by Audibert et al. (2009). The resulting algorithm often outperforms UCB1 and can be shown to be essentially unimprovable. The algorithm that we will describe in Section 3.2.4 implements the OFU principle in MDPs in a way similar to UCB1.

The setting considered here is called the frequentist agnostic setting, where the only assumption made about the distribution of rewards is that they are independent across the actions and time steps and that they belong to the [0, 1] interval. However, there is no other *a priori* knowledge about their distributions. An alternative, historically significant, variant of the problem is when the reward distributions have some known parametric form and the parameters are assumed to be drawn form a known prior distribution. The problem then is to find a policy which maximizes the total expected cumulated discounted reward, where the expectation is both over the random rewards and the parameters of their distributions. This problem can be represented as an MDP whose state at time t is the posterior over the parameters of the reward distributions. For example, if the rewards assume Bernoulli distributions and their parameters are sampled from a Beta distribution then the state at time t will be a $2|\mathcal{A}|$ dimensional vector (since the Beta distribution has two parameters). Thus, the state space of this MDP can be rather complicated even for the simplest examples. In

---

Algorithm 8 The function implementing action selection in UCB1. By assumption, initially $n[a] = 0$, $r[a] = 0$ and the reward received lie in the [0, 1] interval. Further, for $c > 0$, $c/0 = \infty$.

---

function UCB1SELECT(r, n, t)

Input: r, n are arrays of size $|\mathcal{A}|$, t is the number of time steps so far

1: $Umax \leftarrow -\infty$
2: **for all** $a \in \mathcal{A}$ **do**
3:     $U \leftarrow r[a] + \mathcal{R} \cdot \text{sqrt}(2 \cdot \log(t)/n[a])$
4:     **if** $U > Umax$ **then**
5:         $a' \leftarrow a, Umax \leftarrow U$
6:     **end if**
7: **end for**
8: **return** a'

---

his groundbreaking paper, Gittins (1989) has shown that rather surprisingly, the optimal policy in this MDP assumes a simple index-form, which, in some special cases can be calculated exactly and efficiently (e.g., in the case of Bernoulli reward distributions mentioned above). The conceptual difficulty of this so-called *Bayesian approach* is that although the policy is optimal on the average for a collection of randomly chosen environments, there is no guarantee that the policy will perform well on the individual environments. The appeal of the Bayesian approach, however, is that it is conceptually very simple and the exploration problem is reduced to a computational problem.

Algorithm 9 The function implementing the update routine of UCB1. The update, which updates the action counters and the estimates of the average reward, must be called after each interaction.

---

function UCB1UPDATE(A, R, r, n)

Input: A is the last action selected, R is the associated reward, r, n are arrays of size $|\mathcal{A}|$, t is the number of time steps so far

1: $n[A] \leftarrow n[A] + 1$
2: $r[A] = r[A] + 1.0 / n[A] \cdot (R - r[A])$
3: **return** r, n

3.2.2   ACTIVE LEARNING IN BANDITS

Consider now active learning, still in the case when the MDP has a single state. Let the goal be to find an action with the highest immediate reward given (say) T interactions. Since the rewards received during the course of interaction do not matter, the only reason not to try an action is if it can be seen to be worse than some other action with sufficient certainty. The remaining actions should be tried in the hope of proving that some are suboptimal. A simple way to achieve this is to compute upper and lower confidence bounds for each action:

$$U_t(a) = Q_t(a) + \mathcal{R}\sqrt{\frac{\log(2|\mathcal{A}|T/\delta)}{2t}},$$
$$L_t(a) = Q_t(a) - \mathcal{R}\sqrt{\frac{\log(2|\mathcal{A}|T/\delta)}{2t}},$$

and eliminate an action a if $U_t(a) < \max_{a' \in \mathcal{A}} L_t(a')$. Here $0 < \delta < 1$ is a user chosen parameter which specifies the target confidence with which the algorithm is allowed to fail to return an action with the highest expected reward. Apart from constant factors and using estimated variances in the confidence bounds, this algorithm is unimprovable (Even-Dar et al., 2002; Tsitsiklis and Mannor, 2004; Mnih et al., 2008).

3.2.3   ACTIVE LEARNING IN MARKOV DECISION PROCESSES

There exist only a few theoretical works that consider active learning in MDPs. Deterministic environments have been considered by Thrun (1992) (see also Berman, 1998). It turns out that the bounds given in Thrun (1992) can be significantly improved as follows:[2]

Assume that the MDP $\mathcal{M}$ is deterministic. Then the MDP's transition structure can be recovered in at most $n^2 m$ steps, where $n = |\mathcal{X}|, m = |\mathcal{A}|$ (Ortner, 2008). A procedure that achieves this is as follows: The task is to explore all actions in all states exactly once. Hence, at any time t, given the "known part" of the dynamics, we find the closest state with an unexplored action. In at most $n - 1$ steps, this state is reached and the action chosen is explored. Since there are altogether nm state-action pairs to explore, the total number of time steps needed is $n^2 m$.[3] Given the transition structure, the reward structure can be explored up to accuracy ε with probability $1 - \delta$ after at most $k = \log(nm/\delta)/\varepsilon^2$ visits to all state-action pairs. If it takes $e(\leq n^2 m)$ time-steps to visit all state-action pairs by some exploration policy, then in ke steps the learner will have an ε-accurate model of the environment. Knowing such a model allows the learner to find a policy whose value is $4\gamma\varepsilon/(1 - \gamma)^2$-close to the optimal value in each state (assuming, for simplicity, $\gamma \geq 0.5$). Thus, altogether, to find an overall ε-optimal policy, at most $n^2 m + 4e \log(nm/\delta)/((1 - \gamma)^2 \varepsilon)^2$ steps are needed.

According to the author's knowledge, there exist no works that consider the analogous problem of finding a uniformly almost-optimal strategy in a stochastic MDP. Even-Dar et al. (2002) consider active-learning in finite stochastic MDPs but only under the (strong) assumption that the learner can reset the state of the MDP to an arbitrary state. This way they avoid the challenge of navigating in the unknown MDP.

That this is indeed a major challenge can be seen because there exists MDPs where random exploration takes exponential time in the MDPs' size to visit all parts of the state space.

Consider for example a chain-like MDP with n states, say $\mathcal{X} = \{1, 2, \ldots, n\}$. Let $\mathcal{A} = \{L_1, L_2, R\}$. Actions L_1 and L_2 decrement the state by one, while action R increments it by one. The

---

[2]Curiously, this argument is new.

[3]This bound curiously improves the bound of Thrun (1992). The bound can be shown to be tight in an asymptotic sense.

state is not changed at the boundaries when the action would lead to a state outside of $\mathcal{X}$. The policy that selects actions uniformly at random will need $3(2^n - n - 1)$ steps on average to reach state n from state 1 (Howard, 1960). However, a policy that systematically explores the state space will only need $O(n)$ actions to reach state n from state 1. Assume now that all immediate rewards are zero except at state n where the reward is one. Consider an explore-then-exploit learner that explores the MDP randomly until its estimates are sufficient accurate (e.g., until it visited all state-action pairs sufficiently many times). Clearly, the learner will take exponentially many steps before switching to exploitation and hence the learner will suffer a huge regret. The situation is not much better if the agent uses a simple exploration strategy based on some estimates of the values of the actions.

A problem closely related to active learning (without a reset) was studied by Kearns and Singh (2002). They proposed the E^3-algorithm that explores an unknown (stochastic) MDP and stops when it knows a good policy for the state *just visited*. They proved that in discounted MDPs E^3 needs a polynomial number of interactions and uses poly-resources in the relevant parameters of the problem before it stops. In a follow-up work, Brafman and Tennenholtz (2002) introduced the R-max algorithm which refines the E^3 algorithm and proved similar results. Another refinement of E^3 is due to Domingo (1999) who proposed to use adaptive sampling to increase efficiency when the MDP has many near-deterministic transitions. If the problem is undiscounted, both E^3 and R-max need the knowledge of the so-called ε-mixing time of the MDP to work properly. When this knowledge is not available, the algorithms do not know when to stop (Brafman and Tennenholtz, 2002).

Little is known about the performance of active learning algorithms on practical problems. Some experimental results (for some heuristic algorithms) can be found in the paper by Şimşek and Barto (2006).

3.2.4   ONLINE LEARNING IN MARKOV DECISION PROCESSES

Let us now return to online learning in MDPs. One possible goal then is to minimize *regret*, i.e., the difference of the total reward that would have been achieved by the optimal policy and that of received by the learner. This problem is considered in the first part of this section. Another possible goal is to minimize the number of time steps when the algorithm's future expected return falls short of the optimal expected return by some prespecified amount. This problem is considered in the second part of this section.

Regret minimization and the UCRL2 algorithm   Consider a finite (small) MDP $\mathcal{M} = (\mathcal{X}, \mathcal{A}, \mathcal{P}_0)$. Assume that the random immediate rewards are bound to lie in $[0, 1]$ and, for simplicity, assume that all deterministic (stationary) policies visit all states eventually with probability one, i.e., the MDP is *unichain*.

Under this condition, every policy π gives rise to a recurrent Markov chain on $\mathcal{X}$ and a unique stationary distribution μ_π. Define the long-run average reward of π by

$$\rho^\pi = \sum_{x \in \mathcal{X}} \mu_\pi(x) r(x, \pi(x)).$$

(If we want to emphasize the dependence of the long-run average reward on the MDP $\mathcal{M}$, we will write $\rho^\pi(\mathcal{M})$.) Let ρ^* denote the optimal long-run average reward:

$$\rho^* = \max_{\pi \in \Pi_{\text{stat}}} \rho^\pi.$$

Consider some learning algorithm $\mathcal{A}$ (i.e., $\mathcal{A}$ is a history dependent behavior). Define the *regret* of $\mathcal{A}$ by

$$\mathbf{R}_T^\mathcal{A} = \mathcal{R}_T^\mathcal{A} - T\rho^*,$$

where $\mathcal{R}_T^\mathcal{A} = \sum_{t=0}^{T-1} R_{t+1}$ is the sum of rewards received up to time T while following $\mathcal{A}$. Minimizing the regret is clearly equivalent to maximizing the total reward. Hence, from now on, we consider the problem of minimizing regret. Notice that if $\mathbf{R}_T^\mathcal{A} = o(T)$, i.e., if the rate of growth of the regret is sublinear then the long-term average reward of $\mathcal{A}$ is ρ^*, i.e., $\mathcal{A}$ is *consistent*.

The UCRL2 algorithm described below achieves logarithmic regret. In particular, for any initial state, $T \geq 1, \varepsilon > 0$, with probability $1 - 3\delta$ the regret of UCRL2(δ) satisfies

$$\mathbf{R}_T^{\text{UCRL2}(\delta)} = O(D^2 |\mathcal{X}|^2 |\mathcal{A}| \log(T/\delta)/\varepsilon + \varepsilon T).$$

Here D, the so-called *diameter* of the MDP, is defined as the largest number of steps (on average) it takes to reach some state from some other state in the MDP. Moreover, if the confidence parameter of UCRL2 is set to $\delta = 1/(3T)$ then

$$\mathbb{E}\left[\mathbf{R}_T^{\text{UCRL2}(1/(3T))}\right] = O(D^2 |\mathcal{X}|^2 |\mathcal{A}| \log T / g),$$

where g is the "gap" between the performance of the optimal policy and the second best policy (Auer et al., 2010). One issue with this bound is that the gap g could be very small, in which case, the bound might be vacuous for small values of T. An alternative bound, which is independent of g, takes the form

$$\mathbb{E}\left[\mathbf{R}_T^{\text{UCRL2}(1/(3T))}\right] = O(D |\mathcal{X}| \sqrt{|\mathcal{A}| T \log T})$$

(Auer et al., 2010).

Note that these bounds are vacuous when the MDP has an infinite diameter. This happens if the MDP has some states which are not accessible from some other states, i.e., if the MDP has transient states. The only algorithm that is known to enjoy some regret bounds even when the MDP has transient states is due to Bartlett and Tewari (2009). However, this algorithm requires the *a priori* knowledge of some parameter of the MDP. At present, it is not known if such a knowledge is necessary to achieve a non-trivial bound. The issue with transient states is that it is very costly to distinguish between transient states and states that are just hard to reach.

The UCRL2 algorithm (Algorithm 10) implements the optimism in the face of uncertainty principle: It constructs confidence intervals around the estimates of the transition probabilities and the immediate reward function. These define a set of *plausible* MDPs, C_t. When it comes to compute a policy, UCRL2 finds a model $\mathcal{M}_t^* \in C_t$ and a policy π_t^* that gives (approximately) the highest average reward within this class:

$$\rho^{\pi_t^*}(\mathcal{M}_t^*) \geq \max_{\pi, \mathcal{M} \in C_t} \rho^{\pi}(\mathcal{M}) - 1/\sqrt{t}.$$

Note that an essential element of UCRL2 is that it does not update the policy in each time step but waits until the quality of the statistics available for at least one state-action pair is sufficiently improved. This is implemented in line 6 by the check of the visit-count of the current state-action pair.

A crucial step of the algorithm is the computation of π_t^*. This is done by the procedure OptSolve (cf. Algorithm 11), using undiscounted value iteration over a special MDP. In this MDP, the actions are of the form of a pair (a, p), where $a \in \mathcal{A}$ and p is a plausible next-state distribution given the statistics collected so far at (x, a). The next-state distribution associated to (a, p) is exactly p. Further, the immediate reward at x associated to (a, p) is the highest plausible reward given the local statistics at (x, a).

PAC-MDP algorithms   As mentioned before, an alternative to minimizing the regret is to minimize the number of times the learner's future expected return falls short of the optimal return by a prespecified margin (Kakade, 2003). An online learning algorithm is called PAC-MDP if this measure can be bounded with high probability as a polynomial function of the natural parameters of the MDP and if in each time step polynomially many computational steps are performed. Algorithms that are known to be PAC-MDP include R-max (Brafman and Tennenholtz, 2002; Kakade, 2003),[4] MBIE (Strehl and Littman, 2005), Delayed Q-learning (Strehl et al., 2006), the optimistic-initialization-based algorithm of Szita and Lőrincz (2008), and MorMax by Szita and Szepesvári (2010). Of these, MorMax enjoys the best bound for the number of ε-suboptimal steps, T_ε. According to this bound, with probability $1 - \delta$,

$$T_\varepsilon = \tilde{O}\left(|\mathcal{X}||\mathcal{A}|\left(\frac{V_{\max}}{\varepsilon(1-\gamma)^2}\right)^2 \log\left(\frac{1}{\delta}\right)\right),$$

where $\tilde{O}(\cdot)$ hides terms which are logarithmic in the MDP parameters and $V_{\max}$ is an upper bound on the optimal value function (i.e., $V_{\max} \leq \|r\|_\infty/(1-\gamma)$). One notable feature of this bounds that it scales (log) linearly with the size of the state space. A similar bound is available for Delayed Q-learning (though the dependence of this bound is worse on the other parameters), but no bounds with

---

[4]The published proofs for E^3 (Kearns and Singh, 1998) and R-max concern a slightly different criterion; see the discussion of the previous section. Kakade (2003) proved that (an improved version of) R-max is PAC-MDP. He also proved lower bounds.

Algorithm 10 The UCRL2 algorithm.

---

function UCRL2(δ)

Input: $\delta \in [0, 1]$ is a confidence parameter

1: **for all** $x \in \mathcal{X}$ **do** $\pi[x] \leftarrow a_1$ ▷ Initialize policy

2: $n_2, n_3, r, n_2', n_3', r' \leftarrow 0$ ▷ Initialize arrays

3: $t \leftarrow 1$

4: **repeat**

5:     $A \leftarrow \pi[X]$

6:     **if** $n_2'[X, A] \geq \max(1, n_2[X, A])$ **then**   ▷ Enough new information?

7:         $n_2 \leftarrow n_2 + n_2', n_3 \leftarrow n_3 + n_3', r \leftarrow r + r'$ ▷ Update model

8:         $n_2', n_3', r' \leftarrow 0$

9:         $\pi \leftarrow \text{OptSolve}(n_2, n_3, r, \delta, t)$ ▷ Update policy

10:         $A \leftarrow \pi[X]$

11:     **end if**

12:     $(R, Y) \leftarrow \text{ExecuteInWorld}(A)$ ▷ Execute action in the "world"

13:     $n_2'[X, A] \quad \leftarrow n_2'[X, A] + 1$

14:     $n_3'[X, A, Y] \leftarrow n_3'[X, A, Y] + 1$

15:     $r'[X, A] \quad \leftarrow r'[X, A] + R$

16:     $X \leftarrow Y$

17:     $t \leftarrow t + 1$

18: **until** True

---

this feature are yet available for the other algorithms. The algorithms mentioned here all implement the OFU principle in some way.

The main issue with all these algorithms (including UCRL and its variants) is that they are inherently limited to (small) finite spaces. Larger state-spaces are explicitly considered by Kakade et al. (2003) and Strehl and Littman (2008), who considered restricted classes of MDPs and provided "meta-algorithms" to address the exploration problem. There are two difficulties with these approaches. First, in practice, it may be difficult to verify if the particular problem one is interested in belongs to the said classes of MDPs. Second, the proposed algorithms require black-box MDP solvers. Since solving large MDPs is a difficult problem on its own, the algorithms may be hard to implement.

An alternative to the above techniques is to use a Bayesian approach to address the exploration issue (e.g., Dearden et al., 1998, 1999; Strens, 2000; Poupart et al., 2006; Ross and Pineau, 2008). The pros and contras of this approach are the same as in the case of bandits, the only difference being that the computational challenges multiply.

To the best of our knowledge, the only experimental works that concerns online learning in continuous state MDPs are due to Jong and Stone (2007) and Nouri and Littman (2009). Jong and Stone (2007) proposed a method that can be interpreted as a practical implementation

---

Algorithm 11 Procedure for finding an optimistic policy used by UCRL2.

---

function OPTSOLVE(n_2, n_3, r, δ, t)

Input: n_2, n_3 store counters, r stores total rewards, $\delta \in [0, 1]$ is a confidence parameter

1: $u[\cdot] \leftarrow 0, \pi[\cdot] \leftarrow a_1$ $\triangleright$ Initialize policy

2: **repeat**

3:      $M \leftarrow -\infty, m \leftarrow \infty$

4:      $idx \leftarrow \text{sort}(u)$ $\triangleright u[idx[1]] \geq u[idx[2]] \geq \ldots$

5:      **for all** $x \in \mathcal{X}$ **do**

6:          $u_{\text{new}}[\cdot] \leftarrow -\infty$

7:          **for all** $a \in \mathcal{A}$ **do**

8:              $r \leftarrow r[x, a] / n_2[x, a] + \text{sqrt}(7 \cdot \ln(2 \cdot |\mathcal{X}| \cdot |\mathcal{A}| \cdot t / \delta) / (2 \cdot \max(1, n_2[x, a])))$

9:              $c \leftarrow \text{sqrt}(14 \cdot \ln(2 \cdot |\mathcal{A}| \cdot t / \delta) / \max(1, n_2[x, a]))$

10:            $p[\cdot] \leftarrow n_3[x, a, \cdot] / n_2[x, a]$

11:            $p[idx[1]] \leftarrow \min(1, p[idx[1]] + c/2)$

12:            $j \leftarrow |\mathcal{X}| + 1$

13:            **repeat**

14:              $j \leftarrow j - 1$

15:              $P \leftarrow \text{sum}(p[\cdot]) - p[idx[j]]$

16:              $p[idx[j]] \leftarrow \min(0, 1 - P)$

17:            **until** $P + p[idx[j]] > 1$

18:            $v \leftarrow r + \text{inner\_product}(p[\cdot], u[\cdot])$

19:            **if** $v > u_{\text{new}}$ **then**

20:              $\pi[x] \leftarrow a, u_{\text{new}} \leftarrow v$

21:            **end if**

22:          **end for**

23:          $M \leftarrow \max(M, u_{\text{new}} - u[x]), m \leftarrow \min(m, u_{\text{new}} - u[x])$

24:          $u'[x] \leftarrow u_{\text{new}}$

25:      **end for**

26:      $u \leftarrow u'$

27: **until** $M - m \geq 1 / \text{sqrt}(t)$

28: **return** π

---

of the ideas in Kakade et al. (2003), while Nouri and Littman (2009) experimented with multi-resolution regression trees and fitted Q-iteration. The main message of these works is that explicit exploration control can indeed be beneficial.

Despite the potential huge performance gains that can result from using systematic exploration, current practitioners of reinforcement learning largely neglect the issue of systematic exploration or, at best, use simple heuristics to guide exploration. Certainly, there are some cases when systematic exploration is not needed (e.g., Szepesvári, 1997; Nascimento and Powell, 2009). Further,

---

Algorithm 12 The function implementing the tabular Q-learning algorithm. This function must be called after each transition.

function QLEARNING(X, A, R, Y, Q)

Input: X is the last state, A is the last action, R is the immediate reward received, Y is the next state, Q is the array storing the current action-value function estimate

1: $\delta \leftarrow R + \gamma \cdot \max_{a' \in \mathcal{A}} Q[Y, a'] - Q[X, A]$
2: $Q[X, A] \leftarrow Q[X, A] + \alpha \cdot \delta$
3: **return** Q

---

some simple methods, such as optimistic initialization, might give reasonable performance in practice. Since systematic exploration is hardly possible without a good collection of learning algorithms aimed at learning good policies in an efficient manner, in what follows, we will focus on reviewing the algorithms that might belong to such a collection.

3.3 DIRECT METHODS

In this section, we review algorithms whose aim is to approximate the optimal action-value function Q^* directly. The reviewed algorithms can be thought of as sample-based, approximate versions of value iteration that generate some sequence of action-value functions $(Q_k; k \geq 0)$. The idea is that if Q_k is close to Q^*, the policy that is greedy with respect to Q_k will be close to optimalas shown by the bound (1.16).

The first algorithm that we review is Q-learning by Watkins (1989). We start by describing this algorithm for (small) finite MDPs, which is followed by a description of its various extensions that work even in large MDPs.

3.3.1 Q-LEARNING IN FINITE MDPS

Fix a finite MDP $\mathcal{M} = (\mathcal{X}, \mathcal{A}, \mathcal{P}_0)$ and a discount factor γ. The Q-learning algorithm of Watkins (1989) keeps an estimate $Q_t(x, a)$ of $Q^*(x, a)$ for each state-action pair $(x, a) \in \mathcal{X} \times \mathcal{A}$. Upon observing $(X_t, A_t, R_{t+1}, Y_{t+1})$, the estimates are updated as follows:

$$\delta_{t+1}(Q) = R_{t+1} + \gamma \max_{a' \in \mathcal{A}} Q(Y_{t+1}, a') - Q(X_t, A_t),$$
$$Q_{t+1}(x, a) = Q_t(x, a) + \alpha_t \, \delta_{t+1}(Q_t) \, \mathbb{I}_{\{x = X_t, a = A_t\}}, \quad (x, a) \in \mathcal{X} \times \mathcal{A}. \tag{3.1}$$

Here $A_t \in \mathcal{A}$ and $(Y_{t+1}, R_{t+1}) \sim \mathcal{P}_0(\cdot \mid X_t, A_t)$. When learning from a trajectory, $X_{t+1} = Y_{t+1}$, but this is not necessary for the convergence of the algorithm. Q-learning is an instance of TD learning: the updates are based on the TD-error $\delta_{t+1}(Q_t)$. Algorithm 12 shows the pseudocode of Q-learning.

In stochastic equilibrium, one must have $\mathbb{E}\left[\delta_{t+1}(Q) \mid X_t = x, A_t = a\right] = 0$ for any $(x, a) \in \mathcal{X} \times \mathcal{A}$ that is visited infinitely often. A trivial calculation shows that

$$\mathbb{E}\left[\delta_{t+1}(Q) \,\Big|\, X_t = x, A_t = a\right] = T^*Q\,(x, a) - Q(x, a), \qquad x \in \mathcal{X}, a \in \mathcal{A},$$

where T^* is the Bellman optimality operator defined by (1.15). Hence, under the minimal assumption that every state-action pair is visited infinitely often, in stochastic equilibrium, one must have $T^*Q = Q$. Using Fact 1.5, we see that if the algorithm converges, it must converge to Q^* under the stated condition. The sequence $(Q_t; t \geq 0)$ is indeed known to converge to Q^* when appropriate local learning rates are used (Tsitsiklis, 1994; Jaakkola et al., 1994).[5] The rate of convergence of Q-learning was studied by Szepesvári (1997) in an asymptotic setting and later by Even-Dar and Mansour (2003) in a finite-sample setting.

The key observation that lead to the discovery of Q-learning is that unlike the optimal state values, the optimal action-values can be expressed as expectations (compare Equations (1.13) and (1.15)). This, in turn, allows one to estimate the action-values in an incremental manner.

There exist multi-step versions of Q-learning (e.g., Sutton and Barto, 1998, Section 7.6). However, these are not as appealing (and straightforward) as the multi-step extensions of TD(0) since Q-learning is an inherently off-policy algorithm: the temporal differences underlying Q-learning do not telescope even when $X_{t+1} = Y_{t+1}$.

What policy to follow during learning?   A major attraction of Q-learning is its simplicity and that it allows one to use an arbitrary sampling strategy to generate the training data provided that in the limit, all state-action pairs are updated infinitely often. In a closed-loop situation, the commonly used strategies are to sample the actions following the ε-greedy action selection scheme or the Boltzmann scheme (in the latter case, the probability of selecting action a at time t is chosen to be proportional to $e^{\beta Q_t(X_t, a)}$). With appropriate tuning, one can then achieve asymptotic consistency of the behavior policy (cf., Szepesvári, 1998, Section 5.2 and Singh et al., 2000). However, as discussed in Section 3.2, in closed-loop learning, more systematic exploration might be necessary to achieve reasonable online performance (cf. Section 3.2).

Post-decision states   In many practical problems, a set Z (the set of "post-decision states") smaller than $\mathcal{X} \times \mathcal{A}$ can be identified such that the transition probabilities decompose according to

$$\mathcal{P}(x, a, y) = \mathcal{P}_A(f(x, a), y), \qquad x, y \in \mathcal{X}, a \in \mathcal{A}.$$

Here $f : \mathcal{X} \times \mathcal{A} \to Z$ is some *known* transition function and $\mathcal{P}_A : Z \times \mathcal{X} \to [0, 1]$ is an appropriate probability kernel. Function f determines the *deterministic* "effect" of the actions, while $\mathcal{P}_A$ captures their stochastic effect. Many operations research problems enjoy this structure. For example, in the inventory control problem (Example 1.1), $f(x, a) = (x + a) \wedge M$. Further examples are given by Powell (2007). Note that Sutton and Barto (1998) calls post-decision states "afterstates".

---

[5] Watkins (1989) did not provide a rigorous convergence analysis. Watkins and Dayan (1992) gave a proof for the case when all policies eventually lead to an absorbing state.

If a problem admits post-decision states, learning the immediate reward function (if it is not known) and the so-called post-decision state optimal value function, $V_A^* : Z \to \mathbb{R}$, defined by

$$V_A^*(z) = \sum_{y \in \mathcal{X}} \mathcal{P}_A(z, y)\, V^*(y), \qquad z \in Z,$$

might be both more economical and efficient than learning an action-value function. Update rules and action selection strategies can be derived based on the identity $Q^*(x, a) = r(x, a) + \gamma V_A^*(f(x, a))$, which follows immediately from the definitions.

To see another potential advantage of using post-decision state value functions, assume that we have access to the transition probabilities. In such a case, we might be tempted to approximate the state-value function instead of approximating the action-value function. Then, in order to compute the greedy action (which is necessary in many algorithms), we need to compute $\mathrm{argmax}_{a \in \mathcal{A}}\, r(x, a) + \gamma \sum_{y \in \mathcal{X}} \mathcal{P}(x, a, y) V(y)$. This is a so-called stochastic optimization problem (the modifier "stochastic" refers to that the optimization objective is defined with an expectation). This problem might be computationally challenging when the number of next states is large, and/or the number of actions is large (e.g., if $\mathcal{A}$ is a large or infinite subset of a Euclidean space), and/or $\mathcal{P}$ does not enjoy a nice structure. On the other hand, if one uses a post-decision state value function V_A, then computing a greedy action reduces to finding $\mathrm{argmax}_{a \in \mathcal{A}}\, r(x, a) + \gamma V_A(f(x, a))$. Thus, the expectation is avoided, i.e., no stochastic optimization problem needs to be solved. Further, with a judiciously chosen approximation architecture (such as piecewise linear, concave, separable), the optimization problem might be tractable even for large (or infinite) action spaces. Thus, post-decision state value functions might be advantageous as they allow one to avoid another layer of complexity. Of course, the same applies to using action-value functions, but, as discussed previously, post-decision state value functions may require less storage and potentially require fewer samples to learn than action-value functions. For further details, ideas and examples consult (Powell, 2007).

3.3.2 Q-LEARNING WITH FUNCTION APPROXIMATION

The obvious extension of Q-learning to function approximation with parametric forms ($Q_\theta; \theta \in \mathbb{R}^d$) is

$$\theta_{t+1} = \theta_t + \alpha_t\, \delta_{t+1}(Q_{\theta_t})\, \nabla_\theta Q_{\theta_t}(X_t, A_t).$$

(compare this with (2.5) when $\lambda = 0$). Algorithm 13 shows the pseudocode corresponding to the case when a linear function-approximation method is used, i.e., when $Q_\theta = \theta^\top \varphi$ where $\varphi : \mathcal{X} \times \mathcal{A} \to \mathbb{R}^d$.

Although the above update rule is widely used in practice, little can be said about its convergence properties. In fact, since TD(0) is a special case of this algorithm (when there is only one action for every state), just like TD(0), this update rule will also fail to converge when off-policy sampling or nonlinear function approximation is used (cf. Section 2.2.1). The only known convergence result is due to Melo et al. (2008) who prove convergence under rather restrictive conditions on the sample distribution. More recently, along the line of the recent gradient-like TD algorithms,

---

Algorithm 13 The function implementing the Q-learning algorithm with linear function approximation. This function must be called after each transition.

---

function QLEARNINGLINFAPP(X, A, R, Y, θ)

Input: X is the last state, Y is the next state, R is the immediate reward associated with this transition,
  $\theta \in \mathbb{R}^d$ parameter vector

1: $\delta \leftarrow R + \gamma \cdot \max_{a' \in \mathcal{A}} \theta^\top \varphi[Y, a'] - \theta^\top \varphi[X, A]$

2: $\theta \leftarrow \theta + \alpha \cdot \delta \cdot \varphi[X, A]$

3: **return** θ

Maei et al. (2010b) proposed the greedy gradient Q-learning (greedy GQ) algorithm which lifts the previous restrictive conditions: This new algorithm is guaranteed to converge independently of the sampling distribution. However, since the objective function used in the derivation of this algorithm is non-convex, the algorithm may get stuck in local minima even when used with linear function approximation.

State aggregation   Since the above update rule may fail to converge, it is natural to restrict the value function-approximation method employed and/or modify the update procedure as necessary. In this spirit, let us first consider the case when Q_θ is a state (and action) aggregator (cf. Section 2.2). Then, if $((X_t, A_t); t \geq 0)$ is stationary then the algorithm will behave exactly like tabular Q-learning in an appropriately defined "induced MDP". Hence it will converge to some approximation of the optimal action-value function Q^* (Bertsekas and Tsitsiklis, 1996, Section 6.7.7).

Soft state aggregation   One undesirable property of aggregation is that the value function will not be smooth at the boundaries of the underlying regions. Singh et al. (1995) proposed to address this by a "softened" version of Q-learning. In their algorithm, the approximate action-value function has the form of a linear averager: $Q_\theta(x, a) = \sum_{i=1}^d s_i(x, a)\theta_i$, where $s_i(x, a) \geq 0$ $(i = 1, \ldots, d)$ and $\sum_{i=1}^d s_i(x, a) = 1$. The update rule is modified so that at any time, only one component of the parameter vector θ_t is updated. The component to be updated is selected by randomly drawing an index $I_t \in \{1, \ldots, d\}$ from the multinomial distribution with parameters $(s_1(X_t, A_t), \ldots, s_d(X_t, A_t))$.

Interpolation-based Q-learning   Szepesvári and Smart (2004) proposed a modification of this algorithm, which they call *interpolation based Q-learning* (IBQ-learning). IBQ simultaneously updates all the components of the parameter vector thereby reducing the updates' variance. IBQ-learning can also be viewed as a generalization of Q-learning used with state and action aggregation to interpolators (Tsitsiklis and Van Roy, 1996, Section 8 discusses interpolators in the context of fitted value iteration with known models). The idea is to treat every component θ_i of the parameter vector as a value estimate of some "representative" state-action pair, $(x_i, a_i) \in \mathcal{X} \times \mathcal{A}$ $(i = 1, \ldots, d)$. That is, $(Q_\theta; \theta \in \mathbb{R}^d)$ is chosen such that $Q_\theta(x_i, a_i) = \theta_i$ holds for all $i = 1, \ldots, d$. This makes Q_θ an interpolator (explaining the name of the algorithm). Next, choose the similarity functions $s_i : \mathcal{X} \times \mathcal{A} \to [0, \infty)$. For example, one can use $s_i(x, a) = \exp(-c_1 d_1(x, x_i)^2 - c_2 d_2(a, a_i)^2)$, where $c_1, c_2 > 0$, and d_1, d_2 are appropriate "distance" functions. The update rule of IBQ-learning is as follows:

$$\delta_{t+1,i} = R_{t+1} + \gamma \max_{a' \in \mathcal{A}} Q_{\theta_t}(Y_{t+1}, a') - Q_{\theta_t}(x_i, a_i),$$
$$\theta_{t+1,i} = \theta_{t,i} + \alpha_{t,i}\, \delta_{t+1,i}\, s_i(X_t, A_t),$$
$$i = 1, \dots, d.$$

Each component is updated based on how well it predicts the total future reward and how similar its associated state-action pair is to the state-action pair just visited. If the similarity is small, the impact of the error $\delta_{t+1,i}$ on the change of the component will also be small. The algorithm uses local step-size sequences, $(\alpha_{t,i}; t \geq 0)$, i.e., one step-size for each of the components.

Szepesvári and Smart (2004) prove that this algorithm converges almost surely as long as *(i)* the function-class Q_θ satisfies the above interpolation property and the mapping $\theta \mapsto Q_\theta$ is a *non-expansion* (i.e., $\|Q_\theta - Q_{\theta'}\|_\infty \leq \|\theta - \theta'\|_\infty$ holds for any $\theta, \theta' \in \mathbb{R}^d$); *(ii)* the local step-size sequences $(\alpha_{t,i}; t \geq 0)$ are appropriately chosen and *(iii)* all regions of the state-action space $\mathcal{X} \times \mathcal{A}$ are "sufficiently visited" by $((X_t, A_t); t \geq 0)$. They also provide error bounds on the quality of the action-value function learned. The heart of the analysis is that since $\theta \mapsto Q_\theta$ is a non-expansion, the algorithm implements an incremental approximate version of value iteration, with the underlying operator being a contraction. This is because a non-expansion applied after a contraction or a contraction applied after a non-expansion is a contraction. The idea of using non-expansions has first appeared in the works of Gordon (1995) and Tsitsiklis and Van Roy (1996) in the study of fitted value iteration.

Fitted Q-iteration   Fitted Q-iteration implements fitted value iteration with action-value functions. Given the previous iterate, Q_t, the idea is to form a Monte-Carlo approximation to $(T^* Q_t)(x, a)$ at selected state-action pairs and then regress on the resulting points using one's favorite regression method. Algorithm 14 shows the pseudocode of this method.

It is known that fitted Q-iteration might diverge unless a special regressor is used (Baird, 1995; Boyan and Moore, 1995; Tsitsiklis and Van Roy, 1996). Ormoneit and Sen (2002) suggest to use kernel averaging, while Ernst et al. (2005) suggest using tree based regressors. These are guaranteed to converge (say, if the same data is fed to the algorithm in each iteration) as they implement local averaging and as such results of Gordon (1995); Tsitsiklis and Van Roy (1996) are applicable to them. Riedmiller (2005) reports good empirical results with neural networks, at least when new observations obtained by following a policy greedy with respect to the latest iterate are incrementally added to the set of samples used in the updates. That the sample is changed is essential if no good initial policy is available, i.e., when in the initial sample states which are frequently visited by "good" policies are underrepresented (a theoretical argument for why this is important is given by Van Roy (2006) in the context of state aggregation).

Antos et al. (2007) and Munos and Szepesvári (2008) prove finite-sample performance bounds that apply to a large class of regression methods that use empirical risk minimization over a fixed space $\mathcal{F}$ of candidate action-value functions. Their bounds depend on the *worst-case Bellman*

Algorithm 14 The function implementing one iteration of the fitted Q-iteration algorithm. The function must be called until some criterion of convergence is met. appends the new item n to the list S and returns the modified list. The methods predict and regress are specific to the regression method chosen. The method predict(z, θ) should return the predicted value at the input z given the regression parameters θ, while regress(S), given a list of input-output pairs S, should implement a regression algorithm that solves the regression problem given by S and returns new parameters that can be used in predict.

---

function FITTEDQ(D, θ)
Input: $D = ((X_i, A_i, R_{i+1}, Y_{i+1}); i = 1, \ldots, n)$ is a list of transitions, θ are the regressor parameters

1:    $S \leftarrow []$ ▷ Create empty list
2:    **for** $i = 1 \to n$ **do**
3:        $T \leftarrow R_{i+1} + \max_{a' \in \mathcal{A}} \text{predict}((Y_{i+1}, a'), \theta)$ ▷ Target at (X_i, A_i)
4:        $S \leftarrow \text{append}(S, \langle (X_i, A_i), T \rangle)$
5:    **end for**
6:    $\theta \leftarrow \text{regress}(S)$
7:    **return** θ

---

error of $\mathcal{F}$:

$$e_1^*(\mathcal{F}) = \sup_{Q \in \mathcal{F}} \inf_{Q' \in \mathcal{F}} \left\| Q' - T^* Q \right\|_\mu,$$

where μ is the distribution of state-action pairs in the training sample. That is, $e_1^*(\mathcal{F})$ measures how close $\mathcal{F}$ is to $T^*\mathcal{F} \stackrel{\text{def}}{=} \{T^*Q \mid Q \in \mathcal{F}\}$. The bounds derived have the form of the finite-sample bounds that hold in supervised learning (cf. Equation 2.17), except that the approximation error is measured by $e_1^*(\mathcal{F})$. Note that in the earlier-mentioned counterexamples to the convergence of fitted-value iteration, $e_1^*(\mathcal{F}) = \infty$, suggesting that it is the lack of flexibility of the function approximation method that causes divergence.

3.4 ACTOR-CRITIC METHODS

Actor-critic methods implement generalized policy iteration. Remember that policy iteration works by alternating between a complete policy evaluation and a complete policy improvement step. When using sample-based methods or function approximation, exact evaluation of the policies may require infinitely many samples or might be impossible due to the restrictions of the function-approximation technique. Hence, reinforcement learning algorithms simulating policy iteration must change the policy based on incomplete knowledge of the value function.

     Algorithms that update the policy before it is completely evaluated are said to implement *generalized policy iteration* (GPI). In GPI, there are two closely interacting processes of an actor and a critic: the actor aims at improving the current policy, while the critic evaluates the current policy,

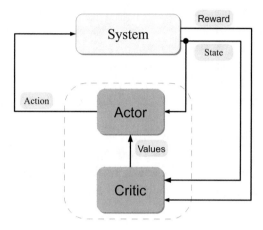

Figure 3.2: The Actor-Critic Architecture

thus helping the actor. The interaction of the actor and the critic is illustrated on Figure 3.2 in a closed-loop learning situation.

Note that, in general, the policy that is used to generate the samples (i.e., the behavior policy) could be different from the one that is evaluated and improved in the actor-critic system (i.e., the target policy). This can be useful because the critic must learn about actions not preferred by the current target policy so that the critic can improve the target policy. This is impossible to achieve if the behavior policy is the same as the target policy *and* if the target policy is deterministic. This is one reason the target policy is usually a stochastic policy. However, even if the target policy is stochastic, the quality of the estimates of the values of low-probability actions can be very poor since less information is received for such actions. It might appear then that choosing actions completely at random might give the most information. However, this is clearly not the case since such a random policy might not visit the important parts of the state-space, as discussed before. Therefore, in practice, the behavior policy often mixes a certain (small) amount of exploration into the target policy.

There are many ways to implement an actor-critic architecture. If the action-space is small, the critic may, e.g., use an approximate action-value function and the actor could follow an ε-greedy or Boltzmann exploration strategy. If the action-space is large or continuous, the actor itself may use function-approximation.

Note that unlike perfect policy iteration, a GPI method may generate a policy that is substantially worse than the previous one. Thus, the quality of the sequence of generated policies may oscillate or even diverge when the policy evaluation step is incomplete, irrespective of whether policy improvement is exact or approximate (Bertsekas and Tsitsiklis, 1996, Example 6.4, p. 283). In practice, GPI tends to generate policies that improve at the beginning. However, at later stages, the policies often oscillate. A common practice, therefore, is to store the sequence of policies obtained and when learning is over, measure the performance of the stored policies by running some tests and then select the empirically best performing one.

Just like in the case of fitted value iteration, the performance of actor-critic methods can be controlled by increasing the "flexibility" of the function approximation methods. Finite-sample performance bounds are given by Antos et al. (2007) when both the actor and the critic use function approximation.

In the next section (Section 3.4.1), we first describe value estimation methods (used by the critic), while in Section 3.4.2, we describe some methods that implement policy improvement (used by the actor). In particular, we first describe greedy methods of policy improvement, followed by a somewhat different idea when the actor uses gradient ascent on the performance function defined by a parametric family of policies.

3.4.1 IMPLEMENTING A CRITIC

The job of the critic is to estimate the value of the current target policy of the actor. This is a value prediction problem. Therefore, the critic can use the methods described in Section 2. Since the actor needs action values, the algorithms are typically modified so that they estimate action values directly. When TD(λ) is appropriately extended, the algorithm known as SARSA(λ) is obtained. This is the first algorithm that we describe below. When LSTD(λ) is extended, we get LSTD-Q(λ), which is described next. λ-LSPE could also be extended, but, for the sake of brevity, this extension is not discussed here.

SARSA   In the case of finite (and small) state and action spaces, similarly to Q-learning, SARSA keeps track of the action-value underlying the possible state-action pairs (Rummery and Niranjan, 1994):

$$\delta_{t+1}(Q) = R_{t+1} + \gamma \, Q(Y_{t+1}, A'_{t+1}) - Q(X_t, A_t),$$
$$Q_{t+1}(x, a) = Q_t(x, a) + \alpha_t \, \delta_{t+1}(Q_t) \, \mathbb{I}_{\{x=X_t, a=A_t\}}, \quad (x, a) \in \mathcal{X} \times \mathcal{A}. \tag{3.2}$$

Here $(Y_{t+1}, R_{t+1}) \sim \mathcal{P}_0(\cdot \mid X_t, A_t)$ and $A'_{t+1} \sim \pi(\cdot | Y_{t+1})$. Compared to Q-learning, the difference is in the definition of the TD error. The algorithm got its name from its use of the current **S**tate, current **A**ction, next **R**eward, next **S**tate, and next **A**ction. When π is fixed, SARSA is just TD(0) applied to state-action pairs. Hence, its convergence follows from the convergence results underlying TD(0).

The multi-step extension of SARSA follows along the lines of the similar extension of TD(0), giving rise to the SARSA(λ) algorithm due to Rummery and Niranjan (1994); Rummery (1995). The extension of the tabular algorithms to the case of function approximation follows along the same lines as the extension of TD(λ). Algorithm 15 shows the pseudocode of SARSA(λ) when it is used with linear function approximation. Being a TD-algorithm, the resulting algorithm is subject to the same limitations as TD(λ) (cf. Section 2.2.1), i.e., it might diverge in off-policy situations. It is, however, possible to extend GTD2 and TDC to work with action values (and use $\lambda > 0$) so that the

---

Algorithm 15 The function implementing the SARSA(λ) algorithm with linear function approximation. This function must be called after each transition.

function SARSALAMBDALINFAPP($X, A, R, Y, A', \theta, z$)

Input: X is the last state, A is the last action chosen, R is the immediate reward received when transitioning to Y, where action A' is chosen. $\theta \in \mathbb{R}^d$ is the parameter vector of the linear function approximation, $z \in \mathbb{R}^d$ is the vector of eligibility traces

1: $\delta \leftarrow R + \gamma \cdot \theta^\top \varphi[Y, A'] - \theta^\top \varphi[X, A]$
2: $z \leftarrow \varphi[X, A] + \gamma \cdot \lambda \cdot z$
3: $\theta \leftarrow \theta + \alpha \cdot \delta \cdot z$
4: **return** (θ, z)

---

resulting algorithms would become free of these limitations. For details consult (Maei and Sutton, 2010).

LSTD-Q(λ)   When LSTD(λ) is generalized to action-value functions, we get the LSTD-Q(λ) algorithm, which solves (2.14), where now $\varphi_t = \varphi(X_t, A_t)$, $\varphi : \mathcal{X} \times \mathcal{A} \to \mathbb{R}^d$, and

$$\delta_{t+1}(\theta) = R_{t+1} + \gamma V_{t+1} - Q_\theta(X_t, A_t),$$

where, assuming that the policy π to be evaluated is a stochastic policy, V_{t+1} is given by

$$V_{t+1} \;=\; \sum_{a \in \mathcal{A}} \pi(a|Y_{t+1})Q_\theta(Y_{t+1}, a) = \langle \theta, \sum_{a \in \mathcal{A}} \pi(a|Y_{t+1})\varphi(Y_{t+1}, a) \rangle.$$

(For deterministic policies this simplifies to $V_{t+1} = Q_\theta(Y_{t+1}, \pi(Y_{t+1}))$.)

If the action space is large and stochastic policies are considered, evaluating the sums $\sum_{a \in \mathcal{A}} \pi(a|x)\varphi(x, a)$ (or integrals in the case of continuous action spaces) might be infeasible. One possibility then is to sample actions from the policy π: $A'_{t+1} \sim \pi(\cdot|Y_{t+1})$ and use $V_{t+1} = Q_\theta(Y_{t+1}, A'_{t+1})$. When the sample consists of trajectories of π, one may set $A'_{t+1} = A_{t+1}$, which gives rise to "SARSA-like" version of LSTD-Q(λ).

An alternative, which is expected to produce better estimates, is to introduce some state features, $\psi : \mathcal{X} \to \mathbb{R}^d$, restrict φ so that $\sum_{a \in \mathcal{A}} \pi(a|x)\varphi(x, a) = 0$ holds for any state $x \in \mathcal{X}$ and define $Q_\theta(x, a) = \theta^\top (\psi(x) + \varphi(x, a))$. Then $V_\theta(x) = \sum_{a \in \mathcal{A}} \pi(a|x)Q_\theta(x, a) = \theta^\top \psi(x)$. Hence, setting $V_{t+1} = V_\theta(Y_{t+1})$ does not introduce any bias, while it is expected to reduce variance since V_{t+1} does not depend on the randomness of A'_{t+1} (Peters et al., 2003; Peters and Schaal, 2008).[6] We will further discuss this choice in the next section.

The pseudocode of LSTD-Q(λ) is shown as Algorithm 16. Note that just like in the case of LSTD(λ), the inverse in line 9 might not exist. Following standard steps, it is possible to derive a recursive version of LSTD-Q(λ).

---

[6] Peters et al. (2003); Peters and Schaal (2008) consider the special case when the parameters between the approximate state-value function V_θ and the action-value function are not shared.

Algorithm 16 The function implementing the LSTD-Q(λ) algorithm with linear function approximation to evaluate a policy π. Note that if π is a deterministic policy than the sum in line 5 can be replaced by $g \leftarrow \varphi[Y_{t+1}, \pi(Y_{t+1})]$.

---

function LSTDQLAMBDA(D, π)

  Input: $D = ((X_t, A_t, R_{t+1}, Y_{t+1}); t = 0, \ldots, n-1)$ is a list of transitions, π is the stochastic policy to be evaluated

  1:   $A, b, z \leftarrow 0$ $\triangleright A \in \mathbb{R}^{d \times d}, b, z \in \mathbb{R}^d$

  2:   **for** $t = 0$ to $n - 1$ **do**

  3:       $f \leftarrow \varphi[X_t, A_t]$

  4:       $z \leftarrow \gamma \cdot \lambda \cdot z + f$

  5:       $g \leftarrow \text{sum}(\pi(\cdot | Y_{t+1}) \cdot \varphi[Y_{t+1}, \cdot])$

  6:       $A \leftarrow A + z \cdot (f - \gamma \cdot g)^{\top}$

  7:       $b \leftarrow b + R_{t+1} \cdot f$

  8:   **end for**

  9:   $\theta \leftarrow A^{-1} b$

 10:   **return** θ

Finally, we note that the various TD-errors defined in this section can also be used in the SARSA algorithm.

3.4.2   IMPLEMENTING AN ACTOR

Policy improvement can be implemented in two ways: One idea is moving the current policy towards the greedy policy underlying the approximate action-value function obtained from the critic. Another idea is to perform gradient ascent directly on the performance surface underlying a chosen parametric policy class. In the next sections, we describe specific methods that implement these ideas.

Greedy improvements   The closest to policy iteration is to let the critic evaluate the current policy based on a lot of data and then switch to the policy that is greedy with respect to the obtained action-value function. Notice that if the action space is finite, the action choices of the greedy policy can be computed "on the fly" (on an as needed basis), i.e., the greedy policy does not need to be explicitly computed or stored, making it possible to use this algorithm in very large, or infinite state spaces. If the policy is evaluated by LSTD-Q(0), this strategy gives rise to the *LSPI (least-squares policy iteration) algorithm* of Lagoudakis and Parr (2003). The variant that uses LSTD-Q(λ) to evaluate policies with a batch of data is shown as Algorithm 17.

    Finite-sample performance bounds for LSPI and generalizations of it are obtained by Antos et al. (2008). Antos et al. (2007) extend these results to continuous action spaces, where

Algorithm 17 The function implementing the LSPI(λ) algorithm with linear function approximation. In practice, the convergence criterion is often replaced by some other criterion. GreedyPolicy(θ)
should return a function that takes as arguments a pair of the form (x, a) and return 1 for the action
that maximizes $\theta^\top \psi(x, \cdot)$, while it returns 0, otherwise.

function LSPI(D, ε)
Input: $D = ((X_t, A_t, R_{t+1}, Y_{t+1}); t = 0, \ldots, n - 1)$ is a list of transitions, ε is an accuracy parameter
1: $\theta' \leftarrow 0$
2: **repeat**
3:     $\theta \leftarrow \theta'$
4:     $\theta' \leftarrow$ LSTDQLambda(D, GreedyPolicy(θ))
5: **until** $\|\theta - \theta'\| > \varepsilon$
6: **return** θ

given the current action value function Q, the next policy is chosen to maximize

$$\rho_{Q,\pi} = \sum_{x \in \mathcal{X}} \mu(x) \int_{\mathcal{A}} Q(x, a)\, \pi(\mathrm{d}a | x)$$

over a restricted policy class. They argue for the necessity of restricting the policies to prevent
overfitting in this case.

The methods mentioned above switch policies without enforcing continuity. This may be
dangerous when the action-value function estimate of the last policy is inaccurate since if the new
policy is radically different than the previous one, it might be hard for the algorithm to recover from
this "failure". In such cases, incremental changes might work better.

One way to ensure incremental changes is to update the parameters ω of a parametric policy-
class (π_ω; $\omega \in \mathbb{R}^{d_\omega}$) by performing stochastic gradient ascent on ρ_{Q,π_ω} (e.g., Bertsekas and Tsitsiklis,
1996, p. 317; Kakade and Langford, 2002 considers such incremental updates when the policies
are given in a tabular form). An indirect way of performing (approximately) greedy updates is to
choose the target policy to be an ε-greedy policy (or a Boltzmann-policy) corresponding to the
current action-value function. Perkins and Precup (2003) analyze this choice with linear function
approximation and when the behavior and target policies are the same. They prove the following
result: Let Γ be the mapping of action-value functions to policies that defines the policy updates.
Assume that *(i)* the exact TD(0) solution is obtained in each iteration and *(ii)* Γ is globally Lipschitz
with a Lipschitz constant smaller than $c(\mathcal{M})$ and the image space of Γ contains only ε-soft policies
(with some fixed $\varepsilon > 0$). Then the sequence of policies generated by the algorithm converges almost
surely. The Lipschitzness property means that $\|\Gamma Q_1 - \Gamma Q_2\| \leq L\|Q_1 - Q_2\|$ holds for all action-
value functions, where both norms are the (unweighted) 2-norms. The constant $c(\mathcal{M})$ depends on
the MDP $\mathcal{M}$. A policy π is called ε-*soft*, if $\pi(a|x) \geq \varepsilon$ holds for all $x \in \mathcal{X}, a \in \mathcal{A}$. More recently,
Van Roy (2006) obtained non-trivial performance bounds for state aggregation for a similar setting.

The methods discussed so far update the policy quite infrequently. An alternative is to interleave the updates of the policy and the value function. Singh et al. (2000) prove asymptotic consistency when *GLIE* (greedy in the limit with infinite exploration) policies are followed and tabular SARSA(0) is used as the critic.

Policy gradient   In this section, we review policy gradient methods (for a sensitivity-based approach, see Cao, 2007). These methods perform stochastic gradient ascent on the performance surface induced by a smoothly parameterized policy class $\Pi = (\pi_\omega; \omega \in \mathbb{R}^{d_\omega})$ of stochastic stationary policies. When the action space is finite, a popular choice is to use the so-called *Gibbs policies*:

$$\pi_\omega(a|x) = \frac{\exp(\omega^\top \xi(x, a))}{\sum_{a' \in \mathcal{A}} \exp(\omega^\top \xi(x, a'))}, \qquad x \in \mathcal{X}, a \in \mathcal{A}.$$

Here $\xi : \mathcal{X} \times \mathcal{A} \to \mathbb{R}^{d_\omega}$ is an appropriate feature-extraction function. If the action space is a subset of a $d_\mathcal{A}$-dimension Euclidean space, a popular choice is to use Gaussian policies when given some parametric mean $g_\omega(x, a)$ and covariance $\Sigma_\omega(x, a)$ functions; the density specifying the action-selection distribution under ω is defined by

$$\pi_\omega(a|x) = \frac{1}{\sqrt{(2\pi)^{d_\mathcal{A}} \det(\Sigma_\omega(x, a))}} \exp\left(-(a - g_\omega(x, a))^\top \Sigma_\omega^{-1}(x, a) (a - g_\omega(x, a))\right).$$

Care must be taken to ensure that Σ_ω is positive definite. For simplicity, Σ_ω is often taken to be $\Sigma_\omega = \beta I$ with some $\beta > 0$.

Given Π, formally, the problem is to find the value of ω corresponding to the best performing policy:

$$\underset{\omega}{\operatorname{argmax}}\, \rho_\omega = ?$$

Here, the performance, ρ_ω, can be measured by the expected return of policy π_ω, with respect to some initial distribution over the states.[7] The initial distribution can be the stationary distribution underlying the policy chosen, in which case maximizing ρ_ω will be equivalent to maximizing the long-run average reward (Sutton et al., 1999a).

The policy gradient theorem   Assume that the Markov chain resulting from following any policy π_ω is ergodic, regardless of the choice of ω. The question is how to estimate the gradient of ρ_ω.

Let $\psi_\omega : \mathcal{X} \times \mathcal{A} \to \mathbb{R}^{d_\omega}$ be the *score function* underlying π_ω:

$$\psi_\omega(x, a) = \frac{\partial}{\partial \omega} \log \pi_\omega(a|x), \qquad (x, a) \in \mathcal{X} \times \mathcal{A}.$$

For example, in the case of Gibbs policies, the score function takes the form $\psi_\omega(x, a) = \xi(x, a) - \sum_{a' \in \mathcal{A}} \pi_\omega(a'|x)\xi(x, a')$.

---

[7]An overall best policy, as measured by the value function, might not exist within the restricted class Π.

Define

$$G(\omega) = \left(Q^{\pi_\omega}(X, A) - h(X)\right) \psi_\omega(X, A).$$ (3.3)

Here (X, A) is a sample from the stationary state-action distribution underlying policy π_ω, Q^{π_ω} is the action-value function of π_ω and h is an arbitrary bounded function. According to the *policy gradient theorem* (see, e.g., Bhatnagar et al., 2009 and the references therein), $G(\omega)$ is an unbiased estimate of the gradient:

$$\nabla_\omega \rho_\omega = \mathbb{E}\left[G(\omega)\right].$$

Let (X_t, A_t) be a sample from the stationary distribution underlying π_{ω_t}. Then, for $(\beta_t; t \geq 0)$, the update rule

$$
\begin{aligned}
\hat{G}_t &= \left(\hat{Q}_t(X_t, A_t) - h(X_t)\right) \psi_\omega(X_t, A_t), \\
\omega_{t+1} &= \omega_t + \beta_t \, \hat{G}_t,
\end{aligned}
$$ (3.4)

implements stochastic gradient ascent as long as

$$\mathbb{E}\left[\hat{Q}_t(X_t, A_t)\psi_{\omega_t}(X_t, A_t)\right] = \mathbb{E}\left[Q^{\pi_{\omega_t}}(X, A)\psi_{\omega_t}(X_t, A_t)\right].$$ (3.5)

The role of h in (3.4) is to reduce the variance of the gradient estimate $\hat{G}_t$ so as to speed up the rate of convergence of the algorithm. Although a good choice can only gain a constant factor in terms of speeding up convergence, in practice, the gain can be substantial. One choice is $h = V^{\pi_{\omega_t}}$, i.e., the value function underlying policy π_{ω_t}. Although this will not explicitly minimize the variance of $\hat{G}_t$ (nor that of $G(\omega_t)$), it is still expected to reduce the variance compared to using $h = 0$ and is thus generally recommended. Of course, the value function of the current policy will normally not be available, but it must be estimated. This can be done together with constructing an estimator $\hat{Q}_t$, as we shall see soon.

As the update rule (3.4) is an instance of stochastic gradient ascent, the sequence (ω_t) will converge almost surely to some local optimum of ρ_ω, provided that the step-size sequence $(\beta_t; t \geq 0)$ satisfies the RM conditions and the problem is sufficiently regular (in general, though, only convergence to a stationary point of ρ_ω can be proven).

The difficulty in implementing (3.4) is twofold: *(i)* One needs to construct an appropriate estimator $\hat{Q}_t$ (and possibly h); *(ii)* The random variables (X_t, A_t) must come from the stationary distribution of π^{ω_t}. In episodic problems, these difficulties can be addressed by updating the parameters at the end of the episodes, giving rise to Williams' REINFORCE algorithm (Williams, 1987). Note that REINFORCE is a *direct policy search algorithm* as it does not use value functions. It is also a member of the family of likelihood ratio methods Glynn, 1990.

In non-episodic problems, a two-timescale algorithm can be used that constructs on estimator $\hat{Q}_t$ on the faster timescale using an appropriate value-function-estimation method and updates the policy parameters on the slower timescale. We now describe an interesting proposal to implement this, due to Sutton et al. (1999a) and Konda and Tsitsiklis (1999).

Compatible function approximation Assume that a linear-in-the-parameters function approximation is used to estimate $\hat{Q}_t$, but choose the feature-extraction function to be the score function underlying the policy class:

$$Q_\theta(x, a) = \theta^\top \psi_\omega(x, a), \qquad (x, a) \in \mathcal{X} \times \mathcal{A}. \tag{3.6}$$

This choice of the function approximation method is called *compatible* with the policy parameterization. Note that the basis functions depend on ω (as $\omega = \omega_t$ changes, ψ_ω will also change). What is a suitable value of θ for a fixed value ω_t? Substituting Q_θ for $\hat{Q}_t$ in (3.5), we get

$$\mathbb{E}\left[\psi_{\omega_t}(X_t, A_t)\psi_{\omega_t}(X_t, A_t)^\top\right]\theta = \mathbb{E}\left[Q^{\pi_{\omega_t}}(X_t, A_t)\psi_{\omega_t}(X_t, A_t)\right].$$

Define $F_\omega = \mathbb{E}\left[\psi_\omega(X, A)\psi_\omega(X, A)^\top\right]$, $g_\omega = \mathbb{E}[Q^{\pi_\omega}(X, A)\psi_\omega(X, A)]$ and let $\theta_*(\omega)$ be the solution to the linear system of equations

$$F_\omega \theta = g_\omega.$$

When this equation holds, $Q_{\theta_*(\omega_t)}$ satisfies (3.5). Notice that $\theta_*(\omega)$ is the parameter that minimizes the mean-squared error

$$\mathbb{E}\left[(Q_\theta(X, A) - Q^{\pi_\omega}(X, A))^2\right].$$

The above derivations suggest the following closed-loop learning algorithm: *(i)* at any time t policy π_{ω_t} is followed; *(ii)* θ_t is updated on the faster timescale (say) by an appropriate version of SARSA(1) *(iii)* the policy parameters are updated on the slower timescale by

$$\omega_{t+1} = \omega_t + \beta_t \left(Q_{\theta_t}(X_t, A_t) - h(X_t)\right) \psi_{\omega_t}(X_t, A_t). \tag{3.7}$$

Algorithm 18 shows the corresponding pseudocode. Konda and Tsitsiklis (2003) proved that (under some regularity conditions) $\liminf_{t\to\infty} \nabla_\omega \rho_{\omega_t} = 0$ holds almost surely if the average-cost version of SARSA(1) is used to update θ_t. They have also shown that if SARSA(λ) is used and $m_\lambda = \liminf_{t\to\infty} \nabla_\omega \rho_{\omega_t}$ then $\lim_{\lambda\to 1} m_\lambda = 0$.

Natural actor-critic Another possible update rule is

$$\omega_{t+1} = \omega_t + \beta_t \theta_t, \tag{3.8}$$

which defines the *natural actor-critic* (NAC) algorithm (the pseudocode of the resulting algorithm differs from that of Algorithm 18 only in that in line 9 the update of ω should be replaced by $\omega \leftarrow \omega + \beta \cdot \theta$). Assuming that F_ω is positive definite, since $g_\omega = \nabla_\omega \rho_\omega$ and $\theta_*(\omega) = F_\omega^{-1}\nabla_\omega \rho_\omega$, we see that $\theta_*(\omega)^\top \nabla_\omega \rho_\omega = \nabla_\omega \rho_\omega^\top F_\omega^{-1} \nabla_\omega \rho_\omega > 0$ unless $\nabla_\omega \rho_\omega = 0$. This shows that the above algorithm implements a stochastic pseudo-gradient algorithm, and thus it converges under the same conditions as (3.7).

Interestingly, the NAC update result in a faster convergence rate than the previous rule. The reason is that $\theta_*(\omega)$ can be shown to be a so-called *natural gradient* (Amari, 1998) of ρ_ω. This

Algorithm 18 An actor-critic algorithm that uses compatible function approximation and SARSA(1).

---

function SARSAActorCritic

1: $\omega, \theta, z \leftarrow 0$

2: $A \leftarrow a_1$

3: **repeat**

4:      $(R, Y) \leftarrow \text{ExecuteInWorld}(A)$

5:      $A' \leftarrow \text{Draw}(\pi_\omega(Y, \cdot))$

6:      $(\theta, z) \leftarrow \text{SARSALambdaLinFApp}(X, A, R, Y, A', \theta, z)$ $\triangleright$ Use $\lambda = 1$ and $\alpha \gg \beta$

7:      $\psi \leftarrow \frac{\partial}{\partial \omega} \log \pi_\omega(X, A)$

8:      $v \leftarrow \text{sum}(\pi_\omega(Y, \cdot) \cdot \theta^\top \varphi[X, \cdot])$

9:      $\omega \leftarrow \omega + \beta \cdot (\theta^\top \varphi[X, A] - v) \cdot \psi$

10:     $X \leftarrow Y$

11:     $A \leftarrow A'$

12: **until** True

---

was first noted by Kakade (2001). Following a natural gradient means that the algorithm performs gradient ascent directly in a metric space underlying the objects of interest, in this case in the space of stochastic policies (with an appropriate metric), as opposed to performing gradient ascent in the (Euclidean) metric space of the parameters (note that the definition of a gradient is dependent on the metric used). In particular, that $\theta_*(\omega)$ is a natural gradient implies that the actual parameterization becomes irrelevant in the sense that the trajectories underlying the ODE $\dot{\omega} = \theta_*(\omega)$ are invariant to arbitrary smooth equivalent reparameterizations of the policy class ($\pi_\omega; \omega \in \mathbb{R}^{d_\omega}$). In the case of the Gibbs policy class, a non-singular linear transformation of the features is a simple example for such a reparameterization. Because of this invariance property, a natural gradient is said to be *covariant*. It is believed that following a natural gradient generally improves the behavior of gradient ascent methods. This is nicely demonstrated by Kakade (2001) on a simple two-state MDP, where the "normal" gradient is very small in a large part of the parameter space, while the natural gradient behaves in a reasonable manner. Other positive examples were given by Bagnell and Schneider (2003); Peters et al. (2003) and Peters and Schaal (2008).

As to the estimation of $\theta^*(\omega)$, Peters and Schaal (2008) (and earlier Peters et al. (2003)) suggest to use LSTD-Q(λ). In particular, they suggested using both state features and the compatible state-action features as described in Section 3.4.1 (note, however, that only $\lambda = 1$ gives an unbiased estimate of the gradient). Their algorithm keeps the value of ω_t fixed until the parameter θ_t as calculated by LSTD-Q(λ) stabilizes. When this happens ω_t is updated by (3.8) and the internal statistics collected by LSTD-Q(λ) is "discounted" by a discount factor $0 < \beta < 1$. They also observe that the original actor-critic (Barto et al., 1983; Sutton, 1984) when used in a finite MDP with no function approximation implements a NAC update. More recently, Bhatnagar et al. (2009) proposed several two-timescale algorithms and proved the convergence of the policy parameters to a neighborhood of the local maxima of the objective function when the critic uses TD(0)-like updates.

CHAPTER 4

For Further Exploration

Inevitably, due to space constraints, this review must miss a large portion of the reinforcement learning literature.

4.1 FURTHER READING

One topic of particular interest not discussed is efficient sampling-based planning (Kearns et al., 1999; Szepesvári, 2001; Kocsis and Szepesvári, 2006; Chang et al., 2008). The main lesson here is that off-line planning in the worst-case can scale exponentially with the dimensionality of the state space (Chow and Tsitsiklis, 1989), while online planning (i.e., planning for the "current state") can break the curse of dimensionality by amortizing the planning effort over multiple time steps (Rust, 1996; Szepesvári, 2001).

Other topics of interest include the linear programming-based approaches (de Farias and Van Roy, 2003, 2004, 2006), dual dynamic programming (Wang et al., 2008), techniques based on sample average approximation (Shapiro, 2003) such as PEGASUS (Ng and Jordan, 2000), online learning in MDPs with arbitrary reward processes (Even-Dar et al., 2005; Yu et al., 2009; Neu et al., 2010), or learning with (almost) no restrictions in a competitive framework (Hutter, 2004).

Other important topics include learning and acting in partially observed MDPs (for recent developments, see, e.g., Littman et al., 2001; Toussaint et al., 2008; Ross et al., 2008), learning and acting in games or under some other optimization criteria (Littman, 1994; Heger, 1994; Szepesvári and Littman, 1999; Borkar and Meyn, 2002), or the development of hierarchical and multi-time-scale methods (Dieterich, 1998; Sutton et al., 1999b).

4.2 APPLICATIONS

The numerous successful applications of reinforcement learning include (in no particular order) learning in games (e.g., Backgammon (Tesauro, 1994) and Go (Silver et al., 2007)), applications in networking (e.g., packet routing (Boyan and Littman, 1994), channel allocation (Singh and Bertsekas, 1997)), applications to operations research problems (e.g., targeted marketing (Abe et al., 2004), maintenance problems (Gosavi, 2004), job-shop scheduling (Zhang and Dieterich, 1995), elevator control (Crites and Barto, 1996), pricing (Rusmevichientong et al., 2006), vehicle routing (Proper and Tadepalli, 2006), inventory control (Chang et al., 2007), fleet management (Simão et al., 2009)), learning in robotics (e.g., controlling quadrupedales (Kohl and Stone, 2004), humanoid robots (Peters et al., 2003), or helicopters

(Abbeel et al., 2007)), and applications to finance (e.g., option pricing (Tsitsiklis and Van Roy, 1999b, 2001; Yu and Bertsekas, 2007; Li et al., 2009)). For further applications, see the lists at the URLs

- `http://www.cs.ualberta.ca/~szepesva/RESEARCH/RLApplications.html` and

- `http://umichrl.pbworks.com/Successes-of-Reinforcement-Learning`.

4.3   SOFTWARE

There are numerous software packages that support the development and testing of RL algorithms. Perhaps, the most notable of these are the RL-Glue and RL-Library packages. The RL-Glue package available from `http://glue.rl-community.org` is intended for helping to standardize RL experiments. It is a free, language-neutral software package that implements a standardized RL interface (Tanner and White, 2009). The RL-Library (`http://library.rl-community.org`) builds on the top of RL-Glue. Its purpose is to provide trusted implementations of various RL testbeds and algorithms. The most notable other RL software packages are CLSquare,[1] PIQLE,[2] RL Toolbox,[3] JRLF[4] and LibPG.[5] These offer the implementation of a large number of algorithms, testbeds, intuitive visualizations, programming tools, etc. Many of these packages support RL-Glue.

[1]`http://www.ni.uos.de/index.php?id=70`
[2]`http://piqle.sourceforge.net/`
[3]`http://www.igi.tugraz.at/ril-toolbox/`
[4]`http://mykel.kochenderfer.com/?page_id=19`
[5]`http://code.google.com/p/libpgrl/`

APPENDIX A

The Theory of Discounted Markovian Decision Processes

The purpose of this section is to give a short proof of the basic results of the theory of Markovian decision processes. All the results will be worked out for the discounted expected total cost criterion. First, we give a short overview of contraction mappings and Banach's fixed-point theorem. Next, we show how this powerful result can be applied to proof a number of basic results about value functions and optimal policies.

A.1 CONTRACTIONS AND BANACH'S FIXED-POINT THEOREM

We start with some basic definitions that we will need in the rest of this section.

Definition A.1 (Norm). Let V be a vector space over the reals. Then $f : V \to \mathbb{R}_0^+$ is a norm on V provided that the following hold:

1. If $f(v) = 0$ for some $v \in V$ then $v = 0$;

2. For any $v, u \in V$, $f(v + u) \leq f(v) + f(u)$.

A vector space together with a norm is called a *normed vector space*. According to the definition a

norm is a function that assigns a nonnegative number to each vector. This number is often called the "length" or just the "norm" of the vector. The norm of a vector v is often denoted by $\|v\|$.

Example A.2 Here are a few examples of norms over the vector space $V = (\mathbb{R}^d, +, \lambda \cdot)$.

1. ℓ^p norms: For $p \geq 1$,

$$\|v\|_p = \left(\sum_{i=1}^d |v_i|^p \right)^{1/p}.$$

2. ℓ^∞ norms:

$$\|v\|_\infty = \max_{1 \leq i \leq d} |v_i|.$$

3. The weighted variants of these norms are defined as follows:

$$\|v\|_p = \begin{cases} \left(\sum_{i=1}^{d} \frac{|v_i|^p}{w_i} \right)^{1/p}, & \text{if } 1 \leq p < \infty; \\ \max_{1 \leq i \leq d} \frac{|v_i|}{w_i}, & \text{if } p = \infty, \end{cases}$$

where $w_i > 0$.

4. The matrix-weighted 2-norm is defined as follows:

$$\|v\|_P^2 = v^T P v.$$

Here P is a fixed, positive definite matrix.

Similarly, one can define norms over spaces of functions. For example, if V is the vector space of functions over the domain $\mathcal{X}$ which are uniformly bounded then

$$\|f\|_\infty = \sup_{x \in \mathcal{X}} |f(x)|.$$

(A function is called uniformly bounded exactly when $\|f\|_\infty < +\infty$.) We will be interested in the convergence of sequences in normed vector spaces.

Definition A.3 (Convergence in norm). Let $V = (V, \|\cdot\|)$ be a normed vector space. Let $v_n \in V$ be a sequence of vectors ($n \in \mathbb{N}$). The sequence ($v_n; n \geq 0$) is said to converge to the vector v in the norm $\|\cdot\|$ if $\lim_{n \to \infty} \|v_n - v\| = 0$. This will be denoted by $v_n \to_{\|\cdot\|} v$. Note that in a d-dimensional vector space $v_n \to_{\|\cdot\|} v$ is the same as requiring that for each $1 \leq i \leq d$, $v_{n,i} \to v_i$ (here $v_{n,i}$ denotes the ith component of v_n). However, this does not hold for infinite dimensional vector spaces. Take for example $\mathcal{X} = [0, 1]$ and the space of uniformly bounded functions over $\mathcal{X}$. Let

$$f_n(x) = \begin{cases} 1, & \text{if } x < 1/n; \\ 0, & \text{otherwise}. \end{cases}$$

Then $f_n(x) \to 0$ for each x (i.e., f_n converges to $f(x) \equiv 0$ *pointwise*). However, $\|f_n - f\|_\infty = \|f_n\|_\infty = 1 \nrightarrow 0$.

   If we have a sequence of real-numbers ($a_n; n \geq 0$), we can test if the sequence converges *without the knowledge of the limiting value* by verifying if it is a *Cauchy sequence*, i.e., whether $\lim_{n \to \infty} \sup_{m \geq n} |a_n - a_m| = 0$. ('Sequences with vanishing oscillations' is possibly a more descriptive name for Cauchy sequences.) It is a quite notable property of the real numbers that every Cauchy sequence of reals assumes a limit.

   The extension of the concept of Cauchy sequences to normed vector spaces is straightforward:

Definition A.4 Cauchy sequence.   Let $(v_n; n \geq 0)$ be a sequence of vectors of a normed vector-space $V = (V, \| \cdot \|)$. Then v_n is called a Cauchy-sequence if $\lim_{n \to \infty} \sup_{m \geq n} \| v_n - v_m \| = 0$.

Normed vector spaces where all Cauchy sequences are convergent are special: one can find examples of normed vector spaces such that some of the Cauchy sequences in the vector space do not have a limit.

Definition A.5 Completeness.   A normed vector space V is called *complete* if every Cauchy sequence in V is convergent in the norm of the vector space.

To pay tribute to Banach, the great Polish mathematicians of the first half of the 20th century, we have the following definition:

Definition A.6 Banach space.   A complete, normed vector space is called a *Banach space*.

One powerful result in the theory of Banach spaces concerns contraction mappings, or contraction operators. These are special Lipschitzian mappings:

Definition A.7   Let $V = (V, \| \cdot \|)$ be a normed vector space. A mapping $T : V \to V$ is called *L-Lipschitz* if for any $u, v \in V$,

$$\| Tu - Tv \| \leq L \| u - v \|.$$

A mapping T is called a *non-expansion* if it is Lipschitzian with $L \leq 1$. It is called a *contraction* if it is Lipschitzian with $L < 1$. In this case L is called the contraction factor of T and T is called an L-contraction.

Note that if T is Lipschitz, it is also continuous in the sense that if $v_n \to_{\|\cdot\|} v$ then also $T v_n \to_{\|\cdot\|} Tv$. This is because $\| T v_n - Tv \| \leq L \| v_n - v \| \to 0$ as $n \to \infty$.

Definition A.8 Fixed point.   Let $T : V \to V$ be some mapping. The vector $v \in V$ is called a *fixed point* of T if $Tv = v$.

Theorem A.9 Banach's fixed-point theorem.   *Let V be a Banach space and $T : V \to V$ be a contraction mapping. Then T has a unique fixed point. Further, for any $v_0 \in V$, if $v_{n+1} = T v_n$ then $v_n \to_{\|\cdot\|} v$, where v is the unique fixed point of T and the convergence is geometric:*

$$\| v_n - v \| \leq \gamma^n \| v_0 - v \|.$$

Proof. Pick any $v_0 \in V$ and define v_n as in the statement of the theorem. We first demonstrate that (v_n) converges to some vector. Then we will show that this vector is a fixed point of T. Finally, we show that T has a single fixed point. Assume that T is a γ-contraction.

To show that (v_n) converges it suffices to show that (v_n) is a Cauchy sequence (since V is a Banach, i.e., complete normed vector-space). We have

$$
\begin{aligned}
\|v_{n+k} - v_n\| = \|T v_{n-1+k} - T v_{n-1}\| \\
\leq \gamma \|v_{n-1+k} - v_{n-1}\| = \gamma \|T v_{n-2+k} - T v_{n-2}\| \\
\leq \gamma^2 \|v_{n-2+k} - v_{n-2}\|
\end{aligned}
$$

$$
\vdots
$$

$$
\begin{aligned}
\leq \gamma^n \|v_k - v_0\| \\
\leq \gamma^n \left(\|v_k\| + \|v_0\|\right).
\end{aligned}
$$

Now,

$$
\|v_k\| \leq \|v_k - v_{k-1}\| + \|v_{k-1} - v_{k-2}\| + \ldots + \|v_1 - v_0\|
$$

and by the same logic as used before $\|v_i - v_{i-1}\| \leq \gamma^{i-1} \|v_1 - v_0\|$. Hence,

$$
\|v_k\| \leq \left(\gamma^{k-1} + \gamma^{k-2} + \ldots + 1\right) \|v_1 - v_0\| \leq \frac{1}{1-\gamma} \|v_1 - v_0\|.
$$

Thus,

$$
\|v_{n+k} - v_n\| \leq \gamma^n \left(\frac{1}{1-\gamma} \|v_1 - v_0\| + \|v_0\|\right)
$$

and so

$$
\limsup_{\substack{n \to \infty \\ k \geq 0}} \|v_{n+k} - v_n\| = 0,
$$

showing that $(v_n; n \geq 0)$ is indeed a Cauchy sequence. Let v be its limit.

Now, let us go back to the definition of the sequence $(v_n; n \geq 0)$:

$$
v_{n+1} = T v_n.
$$

Taking the limes of both sides, on the one hand, we get that $v_{n+1} \to_{\|\cdot\|} v$. On the other hand, $T v_n \to_{\|\cdot\|} T v$, since T is a contraction, hence it is continuous. Thus, the left-hand side converges to v, while the right-hand side converges to $T v$, while the left and right-hand sides are equal. Therefore, we must have $v = T v$, showing that v is a fixed point of T.

Let us consider the problem of uniqueness of the fixed point of T. Let us assume that v, v' are both fixed points of T. Then, $\|v - v'\| = \|T v - T v'\| \leq \gamma \|v - v'\|$, or $(1 - \gamma)\|v - v'\| \leq 0$. Since a norm takes only nonnegative values and $\gamma < 1$, we get that $\|v - v'\| = 0$. Thus, $v - v' = 0$, or $v = v'$.

Finally,

$$\begin{aligned}
\|v_n - v\| &= \|T v_{n-1} - T v\| \\
&\leq \gamma \|v_{n-1} - v\| = \gamma \|T v_{n-2} - T v\| \\
&\leq \gamma^2 \|v_{n-2} - v\| \\
&\;\;\vdots \\
&\leq \gamma^n \|v_0 - v\|.
\end{aligned}$$

$\square$

A.2   APPLICATION TO MDPS

For the purpose of this section, we define V^* by

$$V^*(x) = \sup_{\pi \in \Pi_{\text{stat}}} V^\pi(x), \quad x \in \mathcal{X}.$$

Thus, $V^*(x)$ is an upper bound on the value that we can achieve by choosing some stationary policy π. Note that if the supremum was taken over the larger class of all policies, we could possibly get a larger function. However, in the case of MDPs considered in this section, these two optimal value functions are actually the same. Although, this is not hard to prove, we omit the proof.

Let $B(\mathcal{X})$ be the space of uniformly bounded functions with domain $\mathcal{X}$:

$$B(\mathcal{X}) = \{ V : \mathcal{X} \to \mathbb{R} : \|V\|_\infty < +\infty \}.$$

In what follows, we will view $B(\mathcal{X})$ as a normed-vector space with the norm $\| \cdot \|_\infty$. It is easy to show that $(B(\mathcal{X}), \| \cdot \|_\infty)$ is complete: If $(V_n; n \geq 0)$ is a Cauchy sequence in it then for any $x \in \mathcal{X}$, $(V_n(x); n \geq 0)$ is also a Cauchy sequence over the reals. Denoting by $V(x)$ the limit of $(V_n(x))$, one can show that $\|V_n - V\|_\infty \to 0$. Vaguely speaking, this holds because $(V_n; n \geq 0)$ is a Cauchy sequence in the norm $\| \cdot \|_\infty$ so the rate of convergence of $V_n(x)$ to $V(x)$ is independent of x.

Pick any stationary policy π. Remember that the Bellman operator underlying π, $T^\pi : B(\mathcal{X}) \to B(\mathcal{X})$, is defined by

$$(T^\pi V)(x) = r(x, \pi(x)) + \gamma \sum_{y \in \mathcal{X}} \mathcal{P}(x, \pi(x), y) V(y), \quad x \in \mathcal{X}.$$

Note that T^π is well-defined: If $U \in B(\mathcal{X})$, then $T^\pi U \in B(\mathcal{X})$ holds true.

It is easy to see that V^π as defined by (1.7) is a fixed point to T^π:

$$\begin{aligned}
V^\pi(x) &= \mathbb{E}[R_1 | X_0 = x] + \gamma \sum_{y \in \mathcal{X}} \mathcal{P}(x, \pi(x), y) \mathbb{E}\left[\sum_{t=0}^{\infty} \gamma^t R_{t+2} | X_1 = y \right] \\
&= T^\pi V(x).
\end{aligned}$$

It is also easy to see that T^π is a contraction in $\|\cdot\|_\infty$:

$$
\begin{aligned}
\|T^\pi U - T^\pi V\|_\infty &= \gamma \sup_{x \in \mathcal{X}} \left| \sum_{y \in \mathcal{X}} \mathcal{P}(x, \pi(x), y)(U(y) - V(y)) \right| \\
&\leq \gamma \sup_{x \in \mathcal{X}} \sum_{y \in \mathcal{X}} \mathcal{P}(x, \pi(x), y) |U(y) - V(y)| \\
&\leq \gamma \sup_{x \in \mathcal{X}} \sum_{y \in \mathcal{X}} \mathcal{P}(x, \pi(x), y) \|U - V\|_\infty \\
&= \gamma \|U - V\|_\infty,
\end{aligned}
$$

where the last line follows from $\sum_{y \in \mathcal{X}} \mathcal{P}(x, \pi(x), y) = 1$.

It follows that in order to find V^π, one can construct the sequence $V_0, T^\pi V_0, (T^\pi)^2 V_0, \ldots$, which, by Banach's fixed-point theorem will converge to V^π at a geometric rate.

Now, recall the definition of the Bellman optimality operator: $T^* : B(\mathcal{X}) \to B(\mathcal{X})$,

$$
(T^*V)(x) = \sup_{a \in \mathcal{A}} \left\{ r(x, a) + \gamma \sum_{y \in \mathcal{X}} \mathcal{P}(x, a, y) V(y) \right\}, \quad x \in \mathcal{X}. \tag{A.1}
$$

Again, T^* is well-defined. We now show that T^* is also a γ-contraction with respect to the supremum norm $\|\cdot\|_\infty$.

To see this first note that

$$
|\sup_{a \in \mathcal{A}} f(a) - \sup_{a \in \mathcal{A}} g(a)| \leq \sup_{a \in \mathcal{A}} |f(a) - g(a)|,
$$

which can be seen using an elementary case analysis. Using this inequality and then proceeding as with the analysis of T^π we get,

$$
\begin{aligned}
\|T^*U - T^*V\|_\infty &\leq \gamma \sup_{(x,a) \in \mathcal{X} \times \mathcal{A}} \sum_{y \in \mathcal{X}} \mathcal{P}(x, a, y) |U(y) - V(y)| \\
&\leq \gamma \sup_{(x,a) \in \mathcal{X} \times \mathcal{A}} \sum_{y \in \mathcal{X}} \mathcal{P}(x, a, y) \|U - V\|_\infty \\
&= \gamma \|U - V\|_\infty,
\end{aligned}
$$

thus proving the statement. Here, the last equality follows by $\sum_{y \in \mathcal{X}} \mathcal{P}(x, a, y) = 1$.

The main result of this section is the following theorem:

Theorem A.10 *Let V be the fixed point of T^* and assume that there is policy π which is greedy w.r.t $V : T^\pi V = T^*V$. Then $V = V^*$ and π is an optimal policy.*

Proof. Pick any stationary policy π. Then $T^\pi \le T^*$ in the sense that for any function $V \in B(\mathcal{X})$, $T^\pi V \le T^* V$ holds ($U \le V$ means that $U(x) \le V(x)$ holds for any $x \in \mathcal{X}$). Thus, $V^\pi = T^\pi V^\pi \le T^* V^\pi$, i.e., $V^\pi \le T^* V^\pi$. Since $T^* U \le T^* V$ follows from $U \le V$, we also have $T^* V^\pi \le (T^*)^2 V^\pi$. Chaining the inequalities, we get $V^\pi \le (T^*)^2 V^\pi$. Continuing this way, we get for all $n \ge 0$ that $V^\pi \le (T^*)^n V^\pi$. Since T^* is a contraction, the right-hand side converges to V, the unique fixed point of T^* (at this stage we cannot know if $V = V^*$ or not). Thus, $V^\pi \le V$. Since π was arbitrary, we get that $V^* \le V$.

Pick now a policy π such that $T^\pi V = T^* V$. Since V is the fixed-point of T^*, we have $T^\pi V = V$. Since T^π has a unique fixed point, V^π, we have $V^\pi = V$, showing that $V^* = V$ and that π is an optimal policy. $\qquad\square$

In the statement of the theorem, we were careful in assuming that a greedy policy w.r.t. V exists. Note that this always holds for finite action spaces, and it will hold for infinite action spaces under some extra (continuity) assumptions.

The following theorem serves as the basis of the policy iteration algorithm:

Theorem A.11 Policy improvement theorem. *Choose some stationary policy π_0 and let π be greedy w.r.t. V^{π_0}: $T^\pi V^{\pi_0} = T^* V^{\pi_0}$. Then $V^\pi \ge V^{\pi_0}$, i.e., π is an improvement upon π_0. In particular, if $T^* V^{\pi_0}(x) > V^{\pi_0}(x)$ for some state x then π strictly improves upon π_0 at x: $V^\pi(x) > V^{\pi_0}(x)$. On the other hand, when $T^* V^{\pi_0} = V^{\pi_0}$ then π_0 is an optimal policy.*

Proof. We have $T^\pi V^{\pi_0} = T^* V^{\pi_0} \ge T^{\pi_0} V^{\pi_0} = V^{\pi_0}$. Applying T^π to both sides, we get $(T^\pi)^2 V^{\pi_0} \ge T^\pi V^{\pi_0} \ge V^{\pi_0}$. Continuing this way, we get that for any $n \ge 0$, $(T^\pi)^n V^{\pi_0} \ge V^{\pi_0}$. Taking the limit of both sides, we get that $V^\pi \ge V^{\pi_0}$.

For the second part, notice that we have $(T^\pi)^n V^{\pi_0}(x) \ge T^* V^{\pi_0}(x) > V^{\pi_0}(x)$. Hence, taking the limit, we have $V^\pi(x) \ge T^* V^{\pi_0}(x) > V^{\pi_0}(x)$.

The third part is proven as follows: Since $T^* V^{\pi_0} = V^{\pi_0}$, V^{π_0} is a fixed point of T^*. Since T^* is a contraction, it has a single fixed point, V. Thus $V = V^{\pi_0}$. But we also know that $V^{\pi_0} \le V^* \le V$. Hence, π_0 must be an optimal policy. $\qquad\square$

The policy iteration procedure generates a sequence of policy $\pi_1, \pi_2, \ldots$ such that π_i is greedy w.r.t. $V^{\pi_{i-1}}$, $i = 1, 2, \ldots$. Let us assume further that when choosing a greedy policy, if no improvement is possible, we keep the previous policy and stop the iteration.

We have the following immediate corollary:

Corollary A.12 *If the MDP is finite, the policy iteration procedure terminates in a finite number of steps and returns an optimal policy. Further, a stationary policy of an MDP is optimal if and only if its value function is a fixed point of T^*.*

Proof. From the previous theorem, we know that the sequence of policies is strictly improving. Since in a finite MDP there are a finite number of policies, the procedure must thus terminate. When the procedure terminates, for the final policy π, we have $TV^{\pi} = T^{\pi}V^{\pi} = V^{\pi}$. Thus, by the last part of the previous theorem, π is an optimal policy.

The second part follows immediately from Theorem A.11. $\qquad\square$

Corollary A.13   *Let V be the unique fixed point of T^{*}. Then any policy that is greedy w.r.t. V is an optimal policy. Further, if there exists an optimal stationary policy π^{*} then $V = V^{*}$ and the policy π^{*} is greedy w.r.t. V^{*}.*

Proof. The first part follows immediately from Theorem A.10.

For the second part, assume that π^{*} is an optimal stationary policy. Hence, $V^{\pi^{*}} = V^{*}$. Thus, $V^{\pi^{*}} = T^{\pi^{*}}V^{\pi^{*}} \leq T^{*}V^{\pi^{*}}$. By the second part of Corollary A.12, we must in fact have $T^{*}V^{\pi^{*}} = V^{\pi^{*}}$. Thus $V^{\pi^{*}} \leq V^{*} \leq V = V^{\pi^{*}}$, i.e., all of them are equal and $T^{\pi^{*}}V^{*} = T^{*}V^{*}$. $\qquad\square$

The second part of this corollary in essence shows that the only policies that are optimal are the ones which are greedy w.r.t. to V^{*}.

Bibliography

(2010). *Proceedings of the 27th Annual International Conference on Machine Learning (ICML 2010)*, ACM International Conference Proceeding Series, New York, NY, USA. ACM. 81, 84, 86

A. Prieditis, S. J. R., editor (1995). *Proceedings of the 12th International Conference on Machine Learning (ICML 1995)*, San Francisco, CA, USA. Morgan Kaufmann. 74, 78

Abbeel, P., Coates, A., Quigley, M., and Ng, A. Y. (2007). An application of reinforcement learning to aerobatic helicopter flight. In Schölkopf et al. (2007), pages 1–8. 64

Abe, N., Verma, N. K., Apté, C., and Schroko, R. (2004). Cross channel optimized marketing by reinforcement learning. In Kim, W., Kohavi, R., Gehrke, J., and DuMouchel, W., editors, *Proceedings of the Tenth ACM SIGKDD International Conference on Knowledge Discovery and Data Mining*, pages 767–772, New York, NY, USA. ACM. 63

Albus, J. S. (1971). A theory of cerebellar function. *Mathematical Biosciences*, 10:25–61. DOI: 10.1016/0025-5564(71)90051-4 19

Albus, J. S. (1981). *Brains, Behavior, and Robotics*. BYTE Books, Subsidiary of McGraw-Hill, Peterborough, New Hampshire. 19

Amari, S. (1998). Natural gradient works efficiently in learning. *Neural Computation*, 10(2):251–276. DOI: 10.1162/089976698300017746 60

Antos, A., Munos, R., and Szepesvári, C. (2007). Fitted Q-iteration in continuous action-space MDPs. In Platt et al. (2008), pages 9–16. 35, 51, 54, 56

Antos, A., Szepesvári, C., and Munos, R. (2008). Learning near-optimal policies with Bellman-residual minimization based fitted policy iteration and a single sample path. *Machine Learning*, 71(1):89–129. Published Online First: 14 Nov, 2007. DOI: 10.1007/s10994-007-5038-2 28, 35, 56

Audibert, J.-Y., Munos, R., and Szepesvári, C. (2009). Exploration-exploitation tradeoff using variance estimates in multi-armed bandits. *Theoretical Computer Science*, 410(19):1876–1902. DOI: 10.1016/j.tcs.2009.01.016 39

Auer, P., Cesa-Bianchi, N., and Fischer, P. (2002). Finite time analysis of the multiarmed bandit problem. *Machine Learning*, 47(2-3):235–256. DOI: 10.1023/A:1013689704352 39

Auer, P., Jaksch, T., and Ortner, R. (2010). Near-optimal regret bounds for reinforcement learning. *Journal of Machine Learning Research*, 11:1563—1600. 43

Bagnell, J. A. and Schneider, J. G. (2003). Covariant policy search. In Gottlob, G. and Walsh, T., editors, *Proceedings of the Eighteenth International Joint Conference on Artificial Intelligence (IJCAI-03)*, pages 1019–1024, San Francisco, CA, USA. Morgan Kaufmann. 61

Baird, L. C. (1995). Residual algorithms: Reinforcement learning with function approximation. In A. Prieditis (1995), pages 30–37. 23, 51

Balakrishna, P., Ganesan, R., Sherry, L., and Levy, B. (2008). Estimating taxi-out times with a reinforcement learning algorithm. In *27th IEEE/AIAA Digital Avionics Systems Conference*, pages 3.D.3–1 – 3.D.3–12. DOI: 10.1109/DASC.2008.4702812 11

Bartlett, P. L. and Tewari, A. (2009). REGAL: A regularization based algorithm for reinforcement learning in weakly communicating MDPs. In *Proceedings of the 25th Annual Conference on Uncertainty in Artificial Intelligence*. 43

Barto, A. G., Sutton, R. S., and Anderson, C. W. (1983). Neuronlike adaptive elements that can solve difficult learning control problems. *IEEE Transactions on Systems, Man, and Cybernetics*, 13:834–846. 61

Beleznay, F., Grőbler, T., and Szepesvári, C. (1999). Comparing value-function estimation algorithms in undiscounted problems. Technical Report TR-99-02, Mindmaker Ltd., Budapest 1121, Konkoly Th. M. u. 29-33, Hungary. 16

Berman, P. (1998). On-line searching and navigation. In Fiat, A. and Woeginger, G., editors, *Online Algorithms: The State of the Art*, chapter 10. Springer, Berlin, Heidelberg. 41

Bertsekas, D. P. (2007a). *Dynamic Programming and Optimal Control*, volume 1. Athena Scientific, Belmont, MA, 3 edition. x, 1

Bertsekas, D. P. (2007b). *Dynamic Programming and Optimal Control*, volume 2. Athena Scientific, Belmont, MA, 3 edition. x, 1, 24

Bertsekas, D. P. (2010). Approximate dynamic programming (online chapter). In *Dynamic Programming and Optimal Control*, volume 2, chapter 6. Athena Scientific, Belmont, MA, 3 edition. x, 23

Bertsekas, D. P., Borkar, V. S., and Nedič, A. (2004). Improved temporal difference methods with linear function approximation. In Si, J., Barto, A. G., Powell, W. B., and Wunsch II, D., editors, *Learning and Approximate Dynamic Programming*, chapter 9, pages 235–257. IEEE Press. 31, 32

Bertsekas, D. P. and Ioffe, S. (1996). Temporal differences-based policy iteration and applications in neuro-dynamic programming. LIDS-P-2349, MIT. 30, 32

Bertsekas, D. P. and Shreve, S. (1978). *Stochastic Optimal Control (The Discrete Time Case)*. Academic Press, New York. 1

Bertsekas, D. P. and Tsitsiklis, J. N. (1996). *Neuro-Dynamic Programming*. Athena Scientific, Belmont, MA. x, 17, 23, 24, 50, 53, 57

Bhatnagar, S., Sutton, R. S., Ghavamzadeh, M., and Lee, M. (2009). Natural actor-critic algorithms. *Automatica*. in press. DOI: 10.1016/j.automatica.2009.07.008 59, 61

Borkar, V. S. (1997). Stochastic approximation with two time scales. *Systems & Control Letters*, 29(5):291–294. DOI: 10.1016/S0167-6911(97)90015-3 27

Borkar, V. S. (1998). Asynchronous stochastic approximations. *SIAM J. Control and Optimization*, 36(3):840–851. DOI: 10.1137/S0363012995282784 13

Borkar, V. S. (2008). *Stochastic Approximation: A Dynamical Systems Viewpoint*. Cambridge University Press. 27

Borkar, V. S. and Meyn, S. P. (2002). Risk-sensitive optimal control for Markov decision processes with monotone cost. *Mathematics of Operations Research*, 27(1):192—209. DOI: 10.1287/moor.27.1.192.334 63

Bottou, L. and Bousquet, O. (2008). The trade offs of large scale learning. In Platt et al. (2008), pages 161–168. 33

Boyan, J. A. (2002). Technical update: Least-squares temporal difference learning. *Machine Learning*, 49:233–246. DOI: 10.1023/A:1017936530646 28, 30

Boyan, J. A. and Littman, M. L. (1994). Packet routing in dynamically changing networks: A reinforcement learning approach. In Cowan, J. D., Tesauro, G., and Alspector, J., editors, *NIPS-6: Advances in Neural Information Processing Systems: Proceedings of the 1993 Conference*, pages 671–678. Morgan Kauffman, San Francisco, CA, USA. 63

Boyan, J. A. and Moore, A. W. (1995). Generalization in reinforcement learning: Safely approximating the value function. In Tesauro et al. (1995), pages 369–376. 23, 51

Bradtke, S. J. (1994). *Incremental Dynamic Programming for On-line Adaptive Optimal Control*. PhD thesis, Department of Computer and Information Science, University of Massachusetts, Amherst, Massachusetts. 27

Bradtke, S. J. and Barto, A. G. (1996). Linear least-squares algorithms for temporal difference learning. *Machine Learning*, 22:33–57. DOI: 10.1007/BF00114723 28, 30

Brafman, R. I. and Tennenholtz, M. (2002). R-MAX - a general polynomial time algorithm for near-optimal reinforcement learning. *Journal of Machine Learning Research*, 3:213–231. DOI: 10.1162/153244303765208377 42, 44

Busoniu, L., Babuska, R., Schutter, B., and Ernst, D. (2010). *Reinforcement Learning and Dynamic Programming Using Function Approximators*. Automation and Control Engineering Series. CRC Press. x

Cao, X. R. (2007). *Stochastic Learning and Optimization: A Sensitivity-Based Approach*. Springer, New York. x, 58

Chang, H. S., Fu, M. C., Hu, J., and Marcus, S. I. (2007). An asymptotically efficient simulation-based algorithm for finite horizon stochastic dynamic programming. *IEEE Transactions on Automatic Control*, 52(1):89–94. 63

Chang, H. S., Fu, M. C., Hu, J., and Marcus, S. I. (2008). *Simulation-based Algorithms for Markov Decision Processes*. Springer Verlag. x, 63

Chow, C. S. and Tsitsiklis, J. N. (1989). The complexity of dynamic programming. *Journal of Complexity*, 5:466–488. DOI: 10.1016/0885-064X(89)90021-6 63

Cohen, W. W. and Hirsh, H., editors (1994). *Proceedings of the 11th International Conference on Machine Learning (ICML 1994)*, San Francisco, CA, USA. Morgan Kaufmann. 79, 81

Cohen, W. W., McCallum, A., and Roweis, S. T., editors (2008). *Proceedings of the 25th International Conference Machine Learning (ICML 2008)*, volume 307 of *ACM International Conference Proceeding Series*, New York, NY, USA. ACM. DOI: 10.1145/1390156 78, 81, 82, 86

Cohen, W. W. and Moore, A., editors (2006). *Proceedings of the 23rd International Conference on Machine Learning (ICML 2006)*, volume 148 of *ACM International Conference Proceeding Series*, New York, NY, USA. ACM. DOI: 10.1145/1143844 76, 80, 83, 85

Crites, R. H. and Barto, A. G. (1996). Improving elevator performance using reinforcement learning. In Touretzky, D., Mozer, M. C., and Hasselmo, M. E., editors, *NIPS-8: Advances in Neural Information Processing Systems: Proceedings of the 1995 Conference*, pages 1017–1023, Cambridge, MA, USA. MIT Press. 63

Şimşek, O. and Barto, A. (2006). An intrinsic reward mechanism for efficient exploration. In Cohen and Moore (2006), pages 833—840. 42

Danyluk, A. P., Bottou, L., and Littman, M. L., editors (2009). *Proceedings of the 26th Annual International Conference on Machine Learning (ICML 2009)*, volume 382 of *ACM International Conference Proceeding Series*, New York, NY, USA. ACM. DOI: 10.1145/1553374 80, 85, 87

Dasgupta, S. and Freund, Y. (2008). Random projection trees and low dimensional manifolds. In Ladner, R. E. and Dwork, C., editors, *40th Annual ACM Symposium on Theory of Computing*, pages 537–546. ACM. 20

de Farias, D. P. and Van Roy, B. (2003). The linear programming approach to approximate dynamic programming. *Operations Research*, 51(6):850–865. DOI: 10.1287/opre.51.6.850.24925 63

de Farias, D. P. and Van Roy, B. (2004). On constraint sampling in the linear programming approach to approximate dynamic programming. *Mathematics of Operations Research*, 29(3):462–478. DOI: 10.1287/moor.1040.0094 63

de Farias, D. P. and Van Roy, B. (2006). A cost-shaping linear program for average-cost approximate dynamic programming with performance guarantees. *Mathematics of Operations Research*, 31(3):597–620. DOI: 10.1287/moor.1060.0208 63

De Raedt, L. and Wrobel, S., editors (2005). *Proceedings of the 22nd International Conference on Machine Learning (ICML 2005)*, volume 119 of *ACM International Conference Proceeding Series*, New York, NY, USA. ACM. 77, 85

Dearden, R., Friedman, N., and Andre, D. (1999). Model based Bayesian exploration. In Laskey, K. and Prade, H., editors, *Proceedings of the Fifteenth Conference on Uncertainty in Artificial Intelligence (UAI'99)*, pages 150–159. Morgan Kaufmann. 45

Dearden, R., Friedman, N., and Russell, S. (1998). Bayesian Q-learning. In *Proceedings of the 15th National Conference on Artificial Intelligence (AAAI)*, pages 761–768. AAAI Press. 45

Dietterich, T. (1998). The MAXQ method for hierarchical reinforcement learning. In Shavlik (1998), pages 118–126. 63

Dietterich, T. G., Becker, S., and Ghahramani, Z., editors (2001). *Advances in Neural Information Processing Systems 14*, Cambridge, MA, USA. MIT Press. 79, 81

Domingo, C. (1999). Faster near-optimal reinforcement learning: Adding adaptiveness to the E^3 algorithm. In Watanabe, O. and Yokomori, T., editors, *Proc. of the 10th International Conference on Algorithmic Learning Theory*, volume 1720 of *Lecture Notes in Computer Science*, pages 241–251. Springer. 42

Engel, Y., Mannor, S., and Meir, R. (2005). Reinforcement learning with Gaussian processes. In De Raedt and Wrobel (2005), pages 201–208. DOI: 10.1145/1102351.1102377 36

Ernst, D., Geurts, P., and Wehenkel, L. (2005). Tree-based batch mode reinforcement learning. *Journal of Machine Learning Research*, 6:503–556. 36, 51

Even-Dar, E., Kakade, S. M., and Mansour, Y. (2005). Experts in a Markov decision process. In Saul, L. K., Weiss, Y., and Bottou, L., editors, *Advances in Neural Information Processing Systems 17*, pages 401–408, Cambridge, MA, USA. MIT Press. 63

Even-Dar, E., Mannor, S., and Mansour, Y. (2002). PAC bounds for multi-armed bandit and Markov decision processes. In Kivinen, J. and Sloan, R. H., editors, *Proceedings of the 15th Annual Conference on Computational Learning Theory Computational Learning Theory (COLT 2002)*, volume 2375 of *Lecture Notes in Computer Science*, pages 255–270. Springer. 41

Even-Dar, E. and Mansour, Y. (2003). Learning rates for Q-learning. *Journal of Machine Learning Research*, 5:1–25. 48

Farahmand, A., Ghavamzadeh, M., Szepesvári, C., and Mannor, S. (2008). Regularized fitted Q-iteration: Application to planning. In Girgin, S., Loth, M., Munos, R., Preux, P., and Ryabko, D., editors, *Revised and Selected Papers of the 8th European Workshop on Recent Advances in Reinforcement Learning (EWRL 2008)*, volume 5323 of *Lecture Notes in Computer Science*, pages 55–68. Springer. 36

Farahmand, A., Ghavamzadeh, M., Szepesvári, C., and Mannor, S. (2009). Regularized policy iteration. In Koller et al. (2009), pages 441–448. 36

Frank, J., Mannor, S., and Precup, D. (2008). Reinforcement learning in the presence of rare events. In Cohen et al. (2008), pages 336–343. DOI: 10.1145/1390156.1390199 11

Fürnkranz, J., Scheffer, T., and Spiliopoulou, M., editors (2006). *Proceedings of the 17th European Conference on Machine Learning (ECML-2006)*. Springer. DOI: 10.1007/11871842 80, 83

George, A. P. and Powell, W. B. (2006). Adaptive stepsizes for recursive estimation with applications in approximate dynamic programming. *Machine Learning*, 65:167–198. DOI: 10.1007/s10994-006-8365-9 13, 27

Geramifard, A., Bowling, M. H., Zinkevich, M., and Sutton, R. S. (2007). iLSTD: Eligibility traces and convergence analysis. In Schölkopf et al. (2007), pages 441–448. 33

Ghahramani, Z., editor (2007). *Proceedings of the 24th International Conference on Machine Learning (ICML 2007)*, volume 227 of *ACM International Conference Proceeding Series*, New York, NY, USA. ACM. 78, 82

Ghavamzadeh, M. and Engel, Y. (2007). Bayesian actor-critic algorithms. In Ghahramani (2007), pages 297–304. 36

Gittins, J. C. (1989). *Multi-armed Bandit Allocation Indices*. Wiley-Interscience series in systems and optimization. Wiley, Chichester, NY. 40

Glynn, P. W. (1990). Likelihood ratio gradient estimation for stochastic systems. *Communications of the ACM*, 33(10):75–84. DOI: 10.1145/84537.84552 59

Gordon, G. J. (1995). Stable function approximation in dynamic programming. In A. Prieditis (1995), pages 261–268. 51

Gosavi, A. (2003). *Simulation-based optimization: parametric optimization techniques and reinforcement learning*. Springer Netherlands. x

Gosavi, A. (2004). Reinforcement learning for long-run average cost. *European Journal of Operational Research*, 155(3):654–674. DOI: 10.1016/S0377-2217(02)00874-3 63

Györfi, L., Kohler, M., Krzyżak, A., and Walk, H. (2002). *A distribution-free theory of nonparametric regression*. Springer-Verlag, New York. DOI: 10.1007/b97848 21

Györfi, L., Kohler, M., Krzyżak, A., and Walk, H. (2002). *A Distribution-Free Theory of Nonparametric Regression*. Springer-Verlag. DOI: 10.1007/b97848 34

Härdle, W. (1990). *Applied nonparametric regression*. Cambridge University Press Cambridge. 21

Heger, M. (1994). Consideration of risk in reinforcement learning. In Cohen and Hirsh (1994), pages 105–111. 63

Howard, R. A. (1960). *Dynamic Programming and Markov Processes*. The MIT Press, Cambridge, MA. 42

Hutter, M. (2004). *Universal Artificial Intelligence: Sequential Decisions based on Algorithmic Probability*. Springer, Berlin. 300 pages, http://www.idsia.ch/~marcus/ai/uaibook.htm. 63

Jaakkola, T., Jordan, M., and Singh, S. (1994). On the convergence of stochastic iterative dynamic programming algorithms. *Neural Computation*, 6(6):1185–1201. DOI: 10.1162/neco.1994.6.6.1185 48

Jong, N. K. and Stone, P. (2007). Model-based exploration in continuous state spaces. In Miguel, I. and Ruml, W., editors, *7th International Symposium on Abstraction, Reformulation, and Approximation (SARA 2007)*, volume 4612 of *Lecture Notes in Computer Science*, pages 258–272, Whistler, Canada. Springer. 45

Kaelbling, L., Littman, M., and Moore, A. (1996). Reinforcement learning: A survey. *Journal of Artificial Intelligence Research*, 4:237–285. x

Kakade, S. (2001). A natural policy gradient. In Dietterich et al. (2001), pages 1531–1538. 61

Kakade, S. (2003). *On the sample complexity of reinforcement learning*. PhD thesis, Gatsby Computational Neuroscience Unit, University College London. 44

Kakade, S., Kearns, M. J., and Langford, J. (2003). Exploration in metric state spaces. In Fawcett, T. and Mishra, N., editors, *Proceedings of the 20th International Conference on Machine Learning (ICML 2003)*, pages 306–312. AAAI Press. 45, 46

Kakade, S. and Langford, J. (2002). Approximately optimal approximate reinforcement learning. In Sammut, C. and Hoffmann, A. G., editors, *Proceedings of the 19th International Conference on Machine Learning (ICML 2002)*, pages 267–274, San Francisco, CA, USA. Morgan Kaufmann. 57

Kearns, M. and Singh, S. (2002). Near-optimal reinforcement learning in polynomial time. *Machine Learning*, 49(2–3):209–232. DOI: 10.1023/A:1017984413808 42

Kearns, M. and Singh, S. P. (1998). Near-optimal performance for reinforcement learning in polynomial time. In Shavlik (1998), pages 260–268. DOI: 10.1023/A:1017984413808 44

Kearns, M. J., Mansour, Y., and Ng, A. Y. (1999). Approximate planning in large POMDPs via reusable trajectories. In Solla et al. (1999), pages 1001–1007. 63

Keller, P. W., Mannor, S., and Precup, D. (2006). Automatic basis function construction for approximate dynamic programming and reinforcement learning. In Cohen and Moore (2006), pages 449–456. DOI: 10.1145/1143844.1143901 36

Kocsis, L. and Szepesvári, C. (2006). Bandit based Monte-Carlo planning. In Fürnkranz et al. (2006), pages 282–293. 63

Kohl, N. and Stone, P. (2004). Policy gradient reinforcement learning for fast quadrupedal locomotion. In *Proceedings of the 2004 IEEE International Conference on Robotics and Automation*, pages 2619–2624. IEEE. 63

Koller, D., Schuurmans, D., Bengio, Y., and Bottou, L., editors (2009). *Advances in Neural Information Processing Systems 21*, Cambridge, MA, USA. MIT Press. 78, 82, 86

Kolter, J. Z. and Ng, A. Y. (2009). Regularization and feature selection in least-squares temporal difference learning. In Danyluk et al. (2009), pages 521–528. DOI: 10.1145/1553374.1553442 36

Konda, V. R. and Tsitsiklis, J. N. (1999). Actor-critic algorithms. In Solla et al. (1999), pages 1008–1014. 59

Konda, V. R. and Tsitsiklis, J. N. (2003). On actor-critic algorithms. *SIAM J. Control and Optimization*, 42(4):1143–1166. DOI: 10.1137/S0363012901385691 60

Kosorok, M. R. (2008). *Introduction to Empirical Processes and Semiparametric Inference*. Springer. DOI: 10.1007/978-0-387-74978-5 28

Lagoudakis, M. and Parr, R. (2003). Least-squares policy iteration. *Journal of Machine Learning Research*, 4:1107–1149. DOI: 10.1162/jmlr.2003.4.6.1107 56

Lai, T. L. and Robbins, H. (1985). Asymptotically efficient adaptive allocation rules. *Advances in Applied Mathematics*, 6:4–22. DOI: 10.1016/0196-8858(85)90002-8 39

Lemieux, C. (2009). *Monte Carlo and Quasi-Monte Carlo Sampling*. Springer. 20

Li, Y., Szepesvári, C., and Schuurmans, D. (2009). Learning exercise policies for american options. In *Proc. of the Twelfth International Conference on Artificial Intelligence and Statistics, JMLR: W&CP*, volume 5, pages 352–359. 64

Lin, L.-J. (1992). Self-improving reactive agents based on reinforcement learning, planning and teaching. *Machine Learning*, 9:293–321. DOI: 10.1023/A:1022628806385 27, 32

Littman, M. L. (1994). Markov games as a framework for multi-agent reinforcement learning. In Cohen and Hirsh (1994), pages 157–163. 63

Littman, M. L., Sutton, R. S., and Singh, S. P. (2001). Predictive representations of state. In Dietterich et al. (2001), pages 1555–1561. 63

Maei, H., Szepesvári, C., Bhatnagar, S., Silver, D., Precup, D., and Sutton, R. (2010a). Convergent temporal-difference learning with arbitrary smooth function approximation. In *NIPS-22*, pages 1204–1212. 27

Maei, H., Szepesvári, C., Bhatnagar, S., and Sutton, R. (2010b). Toward off-policy learning control with function approximation. In ICM (2010). 50

Maei, H. R. and Sutton, R. S. (2010). GQ(λ): A general gradient algorithm for temporal-difference prediction learning with eligibility traces. In Baum, E., Hutter, M., and Kitzelmann, E., editors, *Proceedings of the Third Conference on Artificial General Intelligence*, pages 91–96. Atlantis Press. 27, 55

Mahadevan, S. (2009). Learning representation and control in Markov decision processes: New frontiers. *Foundations and Trends in Machine Learning*, 1(4):403–565. DOI: 10.1561/2200000003 36

McAllester, D. A. and Myllymäki, P., editors (2008). *Proceedings of the 24th Conference in Uncertainty in Artificial Intelligence (UAI'08)*. AUAI Press. 83, 86, 87

Melo, F. S., Meyn, S. P., and Ribeiro, M. I. (2008). An analysis of reinforcement learning with function approximation. In Cohen et al. (2008), pages 664–671. DOI: 10.1145/1390156.1390240 49

Menache, I., Mannor, S., and Shimkin, N. (2005). Basis function adaptation in temporal difference reinforcement learning. *Annals of Operations Research*, 134(1):215–238. DOI: 10.1007/s10479-005-5732-z 35

Mnih, V., Szepesvári, C., and Audibert, J.-Y. (2008). Empirical Bernstein stopping. In Cohen et al. (2008), pages 672–679. DOI: 10.1145/1390156.1390241 41

Munos, R. and Szepesvári, C. (2008). Finite-time bounds for fitted value iteration. *Journal of Machine Learning Research*, 9:815–857. 35, 51

Nascimento, J. and Powell, W. (2009). An optimal approximate dynamic programming algorithm for the lagged asset acquisition problem. *Mathematics of Operations Research*, 34:210–237. DOI: 10.1287/moor.1080.0360 46

Nedič, A. and Bertsekas, D. P. (2003). Least squares policy evaluation algorithms with linear function approximation. *Discrete Event Dynamic Systems*, 13(1):79–110. DOI: 10.1023/A:1022192903948 29, 30, 31

Neu, G., György, A., and Szepesvári, C. (2010). The online loop-free stochastic shortest-path problem. In *COLT-10*. 63

Ng, A. Y. and Jordan, M. (2000). PEGASUS: A policy search method for large MDPs and POMDPs. In Boutilier, C. and Goldszmidt, M., editors, *Proceedings of the 16th Conference in Uncertainty in Artificial Intelligence (UAI'00)*, pages 406–415, San Francisco CA. Morgan Kaufmann. 63

Nouri, A. and Littman, M. (2009). Multi-resolution exploration in continuous spaces. In Koller et al. (2009), pages 1209–1216. 45, 46

Ormoneit, D. and Sen, S. (2002). Kernel-based reinforcement learning. *Machine Learning*, 49:161–178. DOI: 10.1023/A:1017928328829 51

Ortner, R. (2008). Online regret bounds for Markov decision processes with deterministic transitions. In Freund, Y., Györfi, L., Turán, G., and Zeugmann, T., editors, *Proc. of the 19th International Conference on Algorithmic Learning Theory (ALT 2008)*, volume 5254 of *Lecture Notes in Computer Science*, pages 123–137. Springer. 41

Parr, R., Li, L., Taylor, G., Painter-Wakefield, C., and Littman, M. L. (2008). An analysis of linear models, linear value-function approximation, and feature selection for reinforcement learning. In Cohen et al. (2008), pages 752–759. DOI: 10.1145/1390156.1390251 24, 25

Parr, R., Painter-Wakefield, C., Li, L., and Littman, M. L. (2007). Analyzing feature generation for value-function approximation. In Ghahramani (2007), pages 737–744. DOI: 10.1145/1273496.1273589 36

Perkins, T. and Precup, D. (2003). A convergent form of approximate policy iteration. In S. Becker, S. T. and Obermayer, K., editors, *Advances in Neural Information Processing Systems 15*, pages 1595–1602, Cambridge, MA, USA. MIT Press. 57

Peters, J. and Schaal, S. (2008). Natural actor-critic. *Neurocomputing*, 71(7–9):1180–1190. DOI: 10.1016/j.neucom.2007.11.026 55, 61

Peters, J., Vijayakumar, S., and Schaal, S. (2003). Reinforcement learning for humanoid robotics. In *Humanoids2003, Third IEEE-RAS International Conference on Humanoid Robots*, pages 225—230. 55, 61, 63

Platt, J. C., Koller, D., Singer, Y., and Roweis, S. T., editors (2008). *Advances in Neural Information Processing Systems 20*, Cambridge, MA, USA. MIT Press. 73, 75, 85, 88

Polyak, B. and Juditsky, A. (1992). Acceleration of stochastic approximation by averaging. *SIAM Journal on Control and Optimization*, 30:838–855. DOI: 10.1137/0330046 13

Poupart, P., Vlassis, N., Hoey, J., and Regan, K. (2006). An analytic solution to discrete Bayesian reinforcement learning. In Cohen and Moore (2006), pages 697–704. DOI: 10.1145/1143844.1143932 45

Powell, W. B. (2007). *Approximate Dynamic Programming: Solving the curses of dimensionality*. John Wiley and Sons, New York. DOI: 10.1002/9780470182963 x, 48, 49

Proper, S. and Tadepalli, P. (2006). Scaling model-based average-reward reinforcement learning for product delivery. In Fürnkranz et al. (2006), pages 735–742. 63

Puterman, M. (1994). *Markov Decision Processes — Discrete Stochastic Dynamic Programming*. John Wiley & Sons, Inc., New York, NY. 1

Rasmussen, C. and Williams, C. (2005). *Gaussian Processes for Machine Learning (Adaptive Computation and Machine Learning)*. The MIT Press. 21

Rasmussen, C. E. and Kuss, M. (2004). Gaussian processes in reinforcement learning. In Thrun, S., Saul, L. K., and Schölkopf, B., editors, *Advances in Neural Information Processing Systems 16*, pages 751–759, Cambridge, MA, USA. MIT Press. 36

Riedmiller, M. (2005). Neural fitted Q iteration – first experiences with a data efficient neural reinforcement learning method. In Gama, J., Camacho, R., Brazdil, P., Jorge, A., and Torgo, L., editors, *Proceedings of the 16th European Conference on Machine Learning (ECML-05)*, volume 3720 of *Lecture Notes in Computer Science*, pages 317–328. Springer. 51

Robbins, H. (1952). Some aspects of the sequential design of experiments. *Bulletin of the American Mathematics Society*, 58:527–535. DOI: 10.1090/S0002-9904-1952-09620-8 38

Ross, S. and Pineau, J. (2008). Model-based Bayesian reinforcement learning in large structured domains. In McAllester and Myllymäki (2008), pages 476–483. 45

Ross, S., Pineau, J., Paquet, S., and Chaib-draa, B. (2008). Online planning algorithms for POMDPs. *Journal of Artificial Intelligence Research*, 32:663–704. 63

Rummery, G. A. (1995). *Problem solving with reinforcement learning*. PhD thesis, Cambridge University. 54

Rummery, G. A. and Niranjan, M. (1994). On-line Q-learning using connectionist systems. Technical Report CUED/F-INFENG/TR 166, Cambridge University Engineering Department. 54

Rusmevichientong, P., Salisbury, J. A., Truss, L. T., Van Roy, B., and Glynn, P. W. (2006). Opportunities and challenges in using online preference data for vehicle pricing: A case study at General Motors. *Journal of Revenue and Pricing Management*, 5(1):45–61. DOI: 10.1057/palgrave.rpm.5160011 63

Rust, J. (1996). Using randomization to break the curse of dimensionality. *Econometrica*, 65:487–516. DOI: 10.2307/2171751 63

Scherrer, B. (2010). Should one compute the temporal difference fix point or minimize the Bellman residual? The unified oblique projection view. In ICM (2010). 24

Schölkopf, B., Platt, J. C., and Hoffman, T., editors (2007). *Advances in Neural Information Processing Systems 19*, Cambridge, MA, USA. MIT Press. 73, 78

Schraudolph, N. (1999). Local gain adaptation in stochastic gradient descent. In *Ninth International Conference on Artificial Neural Networks (ICANN 99)*, volume 2, pages 569–574. DOI: 10.1049/cp:19991170 13

Shapiro, A. (2003). Monte Carlo sampling methods. In *Stochastic Programming, Handbooks in OR & MS*, volume 10. North-Holland Publishing Company, Amsterdam. DOI: 10.1016/S0927-0507(03)10006-0 28, 63

Shavlik, J. W., editor (1998). *Proceedings of the 15th International Conference on Machine Learning (ICML 1998)*, San Francisco, CA, USA. Morgan Kauffmann. 77, 80

Silver, D., Sutton, R. S., and Müller, M. (2007). Reinforcement learning of local shape in the game of Go. In Veloso, M. M., editor, *Proceedings of the 20th International Joint Conference on Artificial Intelligence (IJCAI 2007)*, pages 1053—1058. 32, 63

Simão, H. P., Day, J., George, A. P., Gifford, T., Nienow, J., and Powell, W. B. (2009). An approximate dynamic programming algorithm for large-scale fleet management: A case application. *Transportation Science*, 43(2):178–197. DOI: 10.1287/trsc.1080.0238 63

Singh, S. P. and Bertsekas, D. P. (1997). Reinforcement learning for dynamic channel allocation in cellular telephone systems. In Mozer, M. C., Jordan, M. I., and Petsche, T., editors, *NIPS-9: Advances in Neural Information Processing Systems: Proceedings of the 1996 Conference*, pages 974–980, Cambridge, MA, USA. MIT Press. 63

Singh, S. P., Jaakkola, T., and Jordan, M. I. (1995). Reinforcement learning with soft state aggregation. In Tesauro et al. (1995), pages 361–368. 50

Singh, S. P., Jaakkola, T., Littman, M. L., and Szepesvári, C. (2000). Convergence results for single-step on-policy reinforcement-learning algorithms. *Machine Learning*, 38(3):287–308. DOI: 10.1023/A:1007678930559 48, 58

Singh, S. P. and Sutton, R. S. (1996).   Reinforcement learning with replacing eligibility traces. *Machine Learning*, 32:123–158. DOI: 10.1023/A:1018012322525 17

Singh, S. P. and Yee, R. C. (1994). An upper bound on the loss from approximate optimal-value functions. *Machine Learning*, 16(3):227–233. DOI: 10.1023/A:1022693225949 10

Solla, S. A., Leen, T. K., and Müller, K. R., editors (1999). *Advances in Neural Information Processing Systems 12*, Cambridge, MA, USA. MIT Press. 80, 85

Strehl, A. L., Li, L., Wiewiora, E., Langford, J., and Littman, M. L. (2006).   PAC model-free reinforcement learning.   In Cohen and Moore (2006), pages 881–888. DOI: 10.1145/1143844.1143955 44

Strehl, A. L. and Littman, M. L. (2005). A theoretical analysis of model-based interval estimation. In De Raedt and Wrobel (2005), pages 857–864. DOI: 10.1145/1102351.1102459 44

Strehl, A. L. and Littman, M. L. (2008). Online linear regression and its application to model-based reinforcement learning. In Platt et al. (2008), pages 1417–1424. 45

Strens, M. (2000). A Bayesian framework for reinforcement learning. In Langley, P., editor, *Proceedings of the 17th International Conference on Machine Learning (ICML 2000)*, pages 943–950. Morgan Kaufmann. 45

Sutton, R. S. (1984). *Temporal Credit Assignment in Reinforcement Learning*. PhD thesis, University of Massachusetts, Amherst, MA. 11, 16, 22, 61

Sutton, R. S. (1988). Learning to predict by the method of temporal differences. *Machine Learning*, 3(1):9–44. DOI: 10.1007/BF00115009 11, 16, 22

Sutton, R. S. (1992). Gain adaptation beats least squares. In *Proceedings of the 7th Yale Workshop on Adaptive and Learning Systems*, pages 161—166. 13, 27

Sutton, R. S. and Barto, A. G. (1998). *Reinforcement Learning: An Introduction*. Bradford Book. MIT Press. x, 18, 24, 48

Sutton, R. S., Maei, H. R., Precup, D., Bhatnagar, S., Silver, D., Szepesvári, C., and Wiewiora, E. (2009a). Fast gradient-descent methods for temporal-difference learning with linear function approximation. In Danyluk et al. (2009), pages 993—1000. DOI: 10.1145/1553374.1553501 25, 26, 27

Sutton, R. S., McAllester, D. A., Singh, S. P., and Mansour, Y. (1999a). Policy gradient methods for reinforcement learning with function approximation. In Solla et al. (1999), pages 1057–1063. 58, 59

Sutton, R. S., Precup, D., and Singh, S. P. (1999b). Between MDPs and semi-MDPs: A framework for temporal abstraction in reinforcement learning. *Artificial Intelligence*, 112:181–211. DOI: 10.1016/S0004-3702(99)00052-1 63

Sutton, R. S., Szepesvári, C., Geramifard, A., and Bowling, M. H. (2008). Dyna-style planning with linear function approximation and prioritized sweeping. In McAllester and Myllymäki (2008), pages 528–536. 24

Sutton, R. S., Szepesvári, C., and Maei, H. R. (2009b). A convergent $O(n)$ temporal-difference algorithm for off-policy learning with linear function approximation. In Koller et al. (2009), pages 1609–1616. 25

Szepesvári, C. (1997). The asymptotic convergence-rate of Q-learning. In Jordan, M. I., Kearns, M. J., and Solla, S. A., editors, *Advances in Neural Information Processing Systems 10*, pages 1064–1070, Cambridge, MA, USA. MIT Press. 48

Szepesvári, C. (1997). Learning and exploitation do not conflict under minimax optimality. In Someren, M. and Widmer, G., editors, *Machine Learning: ECML'97 (9th European Conf. on Machine Learning, Proceedings)*, volume 1224 of *Lecture Notes in Artificial Intelligence*, pages 242–249. Springer, Berlin. 46

Szepesvári, C. (1998). *Static and Dynamic Aspects of Optimal Sequential Decision Making*. PhD thesis, Bolyai Institute of Mathematics, University of Szeged, Szeged, Aradi vrt. tere 1, HUNGARY, 6720. 48

Szepesvári, C. (2001). Efficient approximate planning in continuous space Markovian decision problems. *AI Communications*, 13:163–176. 63

Szepesvári, C. and Littman, M. L. (1999). A unified analysis of value-function-based reinforcement-learning algorithms. *Neural Computation*, 11:2017–2059. DOI: 10.1162/089976699300016070 63

Szepesvári, C. and Smart, W. D. (2004). Interpolation-based Q-learning. In Brodley, C. E., editor, *Proceedings of the 21st International Conference on Machine Learning (ICML 2004)*, pages 791–798. ACM. 50, 51

Szita, I. and Lőrincz, A. (2008). The many faces of optimism: a unifying approach. In Cohen et al. (2008), pages 1048–1055. DOI: 10.1145/1390156.1390288 44

Szita, I. and Szepesvári, C. (2010). Model-based reinforcement learning with nearly tight exploration complexity bounds. In ICM (2010). 44

Tadić, V. B. (2004). On the almost sure rate of convergence of linear stochastic approximation algorithms. *IEEE Transactions on Information Theory*, 5(2):401–409. DOI: 10.1109/TIT.2003.821971 14

Tanner, B. and White, A. (2009). RL-Glue: Language-independent software for reinforcement-learning experiments. *Journal of Machine Learning Research*, 10:2133–2136. 64

Taylor, G. and Parr, R. (2009). Kernelized value function approximation for reinforcement learning. In Danyluk et al. (2009), pages 1017–1024. 36

Tesauro, G. (1994). TD-Gammon, a self-teaching backgammon program, achieves master-level play. *Neural Computation*, 6(2):215–219. DOI: 10.1162/neco.1994.6.2.215 63

Tesauro, G., Touretzky, D., and Leen, T., editors (1995). *NIPS-7: Advances in Neural Information Processing Systems: Proceedings of the 1994 Conference*, Cambridge, MA, USA. MIT Press. 75, 84

Thrun, S. B. (1992). Efficient exploration in reinforcement learning. Technical Report CMU-CS-92-102, Carnegie Mellon University, Pittsburgh, PA. 41

Toussaint, M., Charlin, L., and Poupart, P. (2008). Hierarchical POMDP controller optimization by likelihood maximization. In McAllester and Myllymäki (2008), pages 562–570. 63

Tsitsiklis, J. N. (1994). Asynchronous stochastic approximation and Q-learning. *Machine Learning*, 16(3):185–202. DOI: 10.1007/BF00993306 48

Tsitsiklis, J. N. and Mannor, S. (2004). The sample complexity of exploration in the multi-armed bandit problem. *Journal of Machine Learning Research*, 5:623–648. 41

Tsitsiklis, J. N. and Van Roy, B. (1996). Feature-based methods for large scale dynamic programming. *Machine Learning*, 22:59–94. DOI: 10.1023/A:1018008221616 50, 51

Tsitsiklis, J. N. and Van Roy, B. (1997). An analysis of temporal difference learning with function approximation. *IEEE Transactions on Automatic Control*, 42:674–690. DOI: 10.1109/9.580874 23

Tsitsiklis, J. N. and Van Roy, B. (1999a). Average cost temporal-difference learning. *Automatica*, 35(11):1799–1808. DOI: 10.1016/S0005-1098(99)00099-0 24

Tsitsiklis, J. N. and Van Roy, B. (1999b). Optimal stopping of Markov processes: Hilbert space theory, approximation algorithms, and an application to pricing financial derivatives. *IEEE Transactions on Automatic Control*, 44:1840–1851. DOI: 10.1109/9.793723 64

Tsitsiklis, J. N. and Van Roy, B. (2001). Regression methods for pricing complex American-style options. *IEEE Transactions on Neural Networks*, 12:694–703. DOI: 10.1109/72.935083 64

Tsybakov, A. (2009). *Introduction to nonparametric estimation*. Springer Verlag. 21

Van Roy, B. (2006). Performance loss bounds for approximate value iteration with state aggregation. *Mathematics of Operations Research*, 31(2):234–244. DOI: 10.1287/moor.1060.0188 24, 51, 57

Wahba, G. (2003). Reproducing kernel Hilbert spaces – two brief reviews. In *Proceedings of the 13th IFAC Symposium on System Identification*, pages 549–559. 21

Wang, T., Lizotte, D. J., Bowling, M. H., and Schuurmans, D. (2008). Stable dual dynamic programming. In Platt et al. (2008). 63

Watkins, C. J. C. H. (1989). *Learning from Delayed Rewards*. PhD thesis, King's College, Cambridge, UK. 47, 48

Watkins, C. J. C. H. and Dayan, P. (1992). Q-learning. *Machine Learning*, 3(8):279–292. DOI: 10.1023/A:1022676722315 48

Widrow, B. and Stearns, S. (1985). *Adaptive Signal Processing*. Prentice Hall, Englewood Cliffs, NJ. 26, 28

Williams, R. J. (1987). A class of gradient-estimating algorithms for reinforcement learning in neural networks. In *Proceedings of the IEEE First International Conference on Neural Networks*, San Diego, CA. 59

Xu, X., He, H., and Hu, D. (2002). Efficient reinforcement learning using recursive least-squares methods. *Journal of Artificial Intelligence Research*, 16:259–292. 29, 30

Xu, X., Hu, D., and Lu, X. (2007). Kernel-based least squares policy iteration for reinforcement learning. *IEEE Transactions on Neural Networks*, 18:973–992. DOI: 10.1109/TNN.2007.899161 36

Yu, H. and Bertsekas, D. (2007). Q-learning algorithms for optimal stopping based on least squares. In *Proceedings of the European Control Conference*. 64

Yu, J. and Bertsekas, D. P. (2008). New error bounds for approximations from projected linear equations. Technical Report C-2008-43, Department of Computer Science, University of Helsinki. revised July, 2009. 24

Yu, J. Y., Mannor, S., and Shimkin, N. (2009). Markov decision processes with arbitrary reward processes. *Mathematics of Operations Research*. to appear. DOI: 10.1287/moor.1090.0397 63

Zhang, W. and Dietterich, T. G. (1995). A reinforcement learning approach to job-shop scheduling. In Perrault, C. R. and Mellish, C. S., editors, *Proceedings of the Fourteenth International Joint Conference on Artificial Intelligence (IJCAI 95)*, pages 1114–1120, San Francisco, CA, USA. Morgan Kaufmann. 63

Author's Biography

CSABA SZEPESVÁRI

Csaba Szepesvári received his PhD in 1999 from "Jozsef Attila" University, Szeged, Hungary. He is currently an Associate Professor at the Department of Computing Science of the University of Alberta and a principal investigator of the Alberta Ingenuity Center for Machine Learning. Previously, he held a senior researcher position at the Computer and Automation Research Institute of the Hungarian Academy of Sciences, where he headed the Machine Learning Group. Before that, he spent 5 years in the software industry. In 1998, he became the Research Director of Mindmaker, Ltd., working on natural language processing and speech products, while from 2000, he became the Vice President of Research at the Silicon Valley company Mindmaker Inc. He is the coauthor of a book on nonlinear approximate adaptive controllers, published over 80 journal and conference papers and serves as the Associate Editor of IEEE Transactions on Adaptive Control and AI Communications, is on the board of editors of the Journal of Machine Learning Research and the Machine Learning Journal, and is a regular member of the program committee at various machine learning and AI conferences. His areas of expertise include statistical learning theory, reinforcement learning and nonlinear adaptive control.

Printed in the United States
by Baker & Taylor Publisher Services